SOLAR POWER FOR BEGINNERS

THE DIY GUIDE TO EASILY INSTALL A SOLAR POWER SYSTEM IN YOUR HOME

By

Will Smart

© Copyright 2020 By Will Smart

All rights reserved.

This document is geared towards providing exact and reliable information with regards to the topic and issue covered. The publication is sold with the idea that the publisher is not required to render accounting, officially permitted, or otherwise, qualified services. If advice is necessary, legal or professional, a practiced individual in the profession should be ordered.

- From a Declaration of Principles which was accepted and approved equally by a Committee of the American Bar Association and a Committee of Publishers and Associations.

In no way is it legal to reproduce, duplicate, or transmit any part of this document in either electronic means or in printed format. Recording of this publication is strictly prohibited and any storage of this document is not allowed unless with written permission from the publisher. All rights reserved.

The information provided herein is stated to be truthful and consistent, in that any liability, in terms of inattention or otherwise, by any usage or abuse of any policies, processes, or directions contained within is the solitary and utter responsibility of the recipient reader. Under no circumstances will any legal responsibility or blame be held against the publisher for any reparation, damages, or monetary loss due to the information herein, either directly or indirectly.

Respective authors own all copyrights not held by the publisher.

The information herein is offered for informational purposes solely, and is universal as so. The presentation of the information is without contract or any type of guarantee assurance.

The trademarks that are used are without any consent, and the publication of the trademark is without permission or backing by the trademark owner. All trademarks and brands within this book are for clarifying purposes only and are the owned by the owners themselves, not affiliated with this document.

TABLE OF CONTENTS

Introduction --- 5

Introduction To Solar Electricity --- 7

Solar Power For Beginners --- 17

Cheap Solar Energy For Homes --- 24

Make Solar Power For Homes-Show Me How To Build Home Solar Power. --- 26

How Do You Create Your Own, Home-Based Solar Power Generator? --- 28

Solar Power Systems Residential-Create Your Solar Panels And Create Solar Energy --- 30

How To Install Solar Power System - What You Need To Install Solar Power System In Your Home --- 32

Diy Is The Most Cost-Effective Way To Build Solar Power House - 34

Diy Solar Power For Homes --- 36

Solar Heating --- 40

Solar Cooling --- 51

Solar Panels To Power The Growing Global Demand For Air-Conditioners --- 56

Inverter Basics --- 59

Diy Solar Power Kits - Cheap & Easy Way To Use Solar Energy For Homes --- 66

The Guide To The Installation Of Solar Diy Home ------------------ 69

What Equipment Is Required? --- 73

Components For Your Solar Panel System (Photovoltaic) ----------- 78

The Reality Of A Security System ------------------------------------ 86

How To Size A Solar System: A Step-By-Step Walkthrough ------- 89

Eight Tips For Buying Solar Modules For Residential Buildings --- 95

Factors To Consider When Buying Solar Modules ------------------ 100

Solar Panel Installation Manual - Step By Step---------------------- 102

5 Ways To Avoid Short Circuits ------------------------------------- 107

Five Ways To Save Solar Energy ------------------------------------- 111

Energy-Saving Tips That Will Also Save You Money -------------- 114

Conclusion --- 118

INTRODUCTION

Nowadays, solar power is considered one of the cheapest and cleanest sources of free energy. It is a producer of sustainability. Solar energy can now be used in several ways to fuel your houses, saving you money, and making your home much more environmentally friendly. The only thing that you need to be able to make use of this renewable energy is a house guide for solar water.

Solar energy is not only free for homes, it is the purest free energy form available. You save significant amounts of money by installing an environmentally friendly solar energy system for your home. You will need solar power direct to your home to take advantage of this great source of energy.

There are many uses for solar power in your home, including heating your house, warming your swimming pool or providing hot water for your shower. Essentially, to install this, you set up a grid link system and connect your home to the electricity company to take away excess energy. All of this will be learned in the *Home Guidebook for Solar Power*. The power company can send you a check for a refund rather than charging the energy bill every month.

First, in the 1980s, solar panels were used in California, where there were enough solar panels to power more than ten million homes throughout the world. The opportunities for using solar energy are vast and must not be overlooked or forgotten.

Don't underestimate solar energy, the possibilities are massive. Starting in the 1980s, millions of households in California were powered by solar energy and still are today.

Solar energy provides you with resources to fuel your house, to save you money, and to avoid relying on an expensive electricity supply. We are all expected to leave a safe and stable world for our future generations, and solar energy is a strong starting point for home use.

INTRODUCTION TO SOLAR ELECTRICITY

Solar Electricity

When you consider alternative or renewable energy, the first picture is usually large blue or black solar panels on rooftops or mobile road signs with a small panel. These solar panels, also called photovoltaic modules (or photovoltaic modules), turn sunlight into power and, for decades, have become the backbone of renewable energy. This is called the photovoltaic effect, meaning how sunlight is converted to electric power. This approach has been very innovative in extensive implementation but photovoltaics have only been commonly recognized as an alternative means of producing power in recent years.

1958 saw the launch of the first PV module to satellites in orbit. Even now, solar power at the International Space Station is the primary source of electricity. PV is typically used on Earth in places where the use of electricity does not occur, but there is plenty of sunlight. Solar panels are often used for remote purposes; power cabins, RVs, ships, and small electronics, if there is no grid connection. 'Grid-interactive' solar electrical systems have recently begun to gain momentum to integrate solar electricity into our daily lives. Now, while still enjoying the grid's safety net, we can benefit from the available solar energy.

How Solar Panels or PV Modules Work

A solar panel (PV module) is an electricity fluid under the sun in very basic terms. The battery can be charged and can supply regular household electrical devices or 'tracks' with the help of an inverter. In battery-free systems, PV modules may also be used. Aluminum frames, overlaid with tempered glass, and a waterproof back seal is used in most solar panels. The photoreactive cells themselves, often made of silicone, are sandwiched between glass and support materials. There is a junction box on the back of the package, which can have two cables. If no cables exist on the connection box, it can be opened to the electrical terminals where cables can be connected to remove the electricity generated from the module. When cables remain, the box is usually locked and not open to the customer. Sealed boxes are more common.

Solar electricity can be used in many ways. One of the easiest ways is to charge small portable photovoltaic modules such as cell phones and music players. Solar panels can be wired or used to form a solar array individually. There are two major types of electric power systems for residences, cabinets, and offices with larger electricity loads: stand-alone battery-based systems (also known as off-grid systems) and net-based systems (also known as utility-interactive systems). After reading more about them, you will want to determine the best system for your needs.

CELL TECHNOLOGY

Multiple techniques are used to produce solar cells, the panel building blocks. The main types on the market today are:

- Solar Panels Monocrystalline

The manufacturing process, which uses vast amounts of highly processed silicone and much oil, frequently makes monocrystalline solar panels the most costly. At sunlight conversion to electricity, monocrystalline solar cells are about 13-16 percent efficient.

- Solar panels Polycrystalline (aka multi-crystalline)

The cell efficiency of polycrystalline cells is from 11% to 14%, which is why polycrystalline solar panels are slightly lower in cost than monocrystalline solar panels.

- Ribbon solar panels for strings

String-Ribbon solar modules use less silicon than other crystalline types in the production of cells and achieve efficiencies between 12 and 14%.

- Amorphous solar panels (also known as thin films)

The amorphous solar collectors or the amorphous thin film A-si silicon are not made up of single cells but are produced by depositing a light-sensitive compound on a substrate. Although these solar modules have lower efficiencies (generally 7 to 10%), they offer some advantages. They can often be used in warmer climates because they suffer fewer performance losses than other types in warm conditions. Besides, amorphous technology does not use the typical 'glass sandwich' construction, which allows you to create flexible solar modules that are also very resistant.

- CIGS solar panels

The CIGS technology or the di-selenide of copper and indium-gallium does not use silicon at all and can be transformed into plates with or without discrete cells.

There are also hybrid solar modules that use both thin and crystalline technologies to improve energy acquisition. These modules have efficiencies of up to 19%. Researchers are still working on cheaper and more efficient alternatives, but for the foreseeable future, these five types represent what is commercially available.

SOLAR ELECTRICAL PROPERTIES

Photovoltaic modules generate direct current. This is the same type of electricity generated by a car battery or other batteries. The devices in our homes use a different type of electricity (AC) or DC. Direct current flows only in one direction, while alternating currents quickly change direction and offer some advantages during transmission for example; travelling longer distances due to smaller wires.

To use solar electricity for typical household appliances, a solar inverter that converts direct current to an alternating current is required. An inverter is not needed for a small system to recharge batteries or to supply small electrical devices (such as cell phones and personal music players). However, be sure to use the correct adapters and transformers or voltage limiters if necessary.

A basic introduction to electricity will help clarify this:

- Volt - Electrical potential unit
- Ampere - Current flow rate
- Watt - A rate of supply or consumption of energy

Let's now consider these values in terms of energy and performance.:

- Power is the speed at which energy is supplied, such as your car's speedometer, and is measured in watts
- Energy is the measure of performance over time - like your car's odometer, and this is measured in watt-hours or kilowatt-hours

One of the best examples is your old-school light bulb. If the rated power is 60 watts and you leave the light on for an hour, you will use 60 watts of energy.

Consumption can, therefore, be measured in the following formula:

Power (W) * Time (Hrs) = Energy (Watt-hours)

The odometer counts the number of miles journeyed in the car analogy. The kilowatt-hour numbers (kWh or 1,000 watt-hours) you use per month are shown in your electricity bill. How does a solar panel's wattage rating impact the energy produced by the panel? Easy-with everything else equal, a solar panel like higher wattage can produce more electricity over time than a low wattage panel. During the same period, a 100-watt solar panel will generate twice as much energy as a 50-watt solar panel. It should also be pointed out that the wattage performance rate of solar modules is based on the production of modules called Standard Test Conditions (STC) under laboratory-controlled conditions. STC permit the use of the same metric to compare solar panels to each other. Since these rated wattages are ideal for laboratory use, the module may produce fewer wattages in actual use.

There are also PV modules and ratings. A panel is rated by its operating voltage multiplied by the current: watts = voltage x amps. The quantity of energy produced by a panel during watt-hours is the product of the panel wattage and the number of hours of full sunlight that it receives.

The panel produces, for example, 200 watt-hours of power that outputs 100 watts over two hours. Also, STC is linked to and based on location insulation values. They can be found in most places in the USA and the world in data tables. The number of watts a panel produces, as many factors affect the efficiencies of system components, is likely to be less than these theoretical values. Usual considerations for real-life energy losses are used, but the basic electrical concepts are our priority.

VOLTAGE AND SOLAR PANELS

Three specific voltage levels have been developed for PV modules to learn.

- Nominal voltage: A panel's nominal voltage can also be called 'conversational voltage.' We most often use nominal voltage when we speak about panel voltage and other system components. Besides, the nominal tension corresponds to the battery voltage that the module is ideally suited to charge; it is an exception to the rule that solar panels are used for battery charging. When illumination and temperature conditions change the panel's real voltage output, the panel is not working with any particular voltage. Nominal voltage allows us to ensure that the panels are compliant with a specific system at a glance.

The panel's highest voltages can be produced when connecting with a network and operating at peak efficiency.

The maximum voltage that a panel can produce when it is not connected to the power or system is Open Circuit Voltage or Voc. With a meter that contacts the ends of the built-in cables of the panel, Voc can be measured.

CURRENT AND SOLAR PANELS

There are two current ratings in tables too, (Imp) and (Isc), both of which are classified in amps. The maximum power current is equivalent to Vmp. When the panel operates at high efficiency in a circuit, it is the highest power available. Similar to Voc, when your meter is in contact with the positive and negative end pipe of the panel without being connected to a system or load, the short circuit current indicates the current measurement.

The thickness of the PV system and the elements are measured with all these electrical characteristics. The label on every solar module and manufacturer's specifications can be found in these specifications.

SOLAR PANEL OR PV MODULE CONNECTORS

On the back of the PV modules are two main connector types. Solar panels with a rating of fewer than 100 watts or modules produced more than ten years ago, often have opening boxes. The wires can be connected via a knock-out hole in the side of the junction box to the positive and negative terminals on the panel. A separation box with wire pipes, usually two to three feet in length, is another type of connector.

There are few cases where the wires are bare, but most modules produced today are fitted with MC or multi-contact brand connectors on the ends. The two most commonly referred to as MC4 and H4 are MC connectors in several forms.

LOCATION AND ORIENTATION OF THE SOLAR ARRAY

PV panels are located critically for their energy output. Over one year, a Florida-based solar panel produces more energy than a North

Dakota array. Less sunshine (also called irradiance) will become accessible throughout the year in areas close to the equator than in locations farther north or south. The more irradiance a PV panel touches, the more energy it produces.

The directional orientation of the modules is another factor that will influence your array's electrical output. If you settle for a garden spot, you know that southern exposure increases yield. So in solar panels it's the best way to get maximum exposure when the sun goes from east to west, facing the south.

If a compass is used to locate south of the border, ensure that the magnetic decline adjusts according to the position. The array also affects the production of electricity. A tilted angle equal to your latitude is the best product in the northern hemisphere throughout the year. A tilted angle equal to your latitude less than 15 degrees is going to favor summer production, and winter production will benefit from a perspective that is equal to your latitude more than 15 degrees.

SHADING – DON'T LET IT HAPPEN TO YOU!

Shading is one of the major environmental factors influencing the production of solar electricity. PV modules are highly shade-sensitive. For example, a PV module will lose up to 80 percent of its efficiency when shaded by as little as a leafless tree branch. Choose one with the least shade between 9 AM and 3 PM when you choose a site for installation.

Take things into account when the sun rises higher or lower in the sky at different times of the year. A site where there is no shading in June could be shaded in December for a large part of the day. A solar pathfinder is a useful tool for determining a perfect location for solar production. Plan for your PV panels to generate power for up to 20 years and consider young trees may not be a problem now, but may become a source of shade as they grow.

KNOWLEDGE IS POWER

In a solar electrical system, PV modules are only one component. Check some of our other items to learn more about the different elements that your system will need. Remember the danger and lethality of electricity. For information on the implementation of the code-compliant network, check the Section 690 National Electric Code. Read the 2005 National Electro-Code: Recommended Practices for John Wiles's Photovoltaic Power Systems. Please ensure that you contact your local electrical inspector before installation if you plan an extensive system.

SOLAR POWER FOR BEGINNERS

Whether solar panels work is key to specific potential scenarios (and why they're taking over the world). On the micro stage, in America and around the world, a thriving solar industry is thriving. The Solar Power Industries Association (SEIA) has been claiming the industries achieved an annual growth rate of 50% in the last decade since Congress introduced a commercial loan in 2006. That'd be macro news in most areas. But solar energy has a mission to save the world, alongside making profits.

There is no solution to keep man-made global warming from gradually distorting the Earth's atmosphere and the electricity they will harvest, without solar panels. 'Renewable energy solutions have proven their role in mitigating climate change,' says UNDP. Some people in the industry claim that by 2050 solar power will rise 6,500% as a technology to offset this need.

However, solar panels are still enigmatic with all their significance. They don't look or act like a hero, rigid and somewhat threatening looking black rectangles, but, how do they work?

A Short History

Solar energy research began in 1839 with the observation of what is now known as the photovoltaic effect, by a young French physicist called Edmond Becquerel. His brother, Antoine Becquerel, was a well-known French physicist who was more interested in electricity. Becquerel became active with the family business, ande while he was only 19, he discovered that electricity was generated via sunlight. Edmond became fascinated in how light works and his two interests collided.

As years have passed, the technology has taken tiny but steady steps forward. Writers such as Maria Telkes experimented during the 1940s with the use of sodium sulfates to store sunlight to create the Dover Sun House. During the semiconductor investigation, engineer Russell Shoemaker Ochs inspected a broken sample of silicon and discovered that electricity was being transmitted amid the break.

However, the most significant leap came on 25 April 1954, when physicist and engineer Gerald Pearson, chemist Calvin Fuller, and Daryl Chapin revealed that the first practical silicon solar cell had been built.

Unlike Ochs, the pair worked for Bell Labs and had previously taken on the task of maintaining that harmony. Chapin had sought to create power sources in deserts for remote telephones if standard batteries had dried up. Pearson and Fuller focused on the management of semiconductor properties, which would be used later on to drive computer systems. The three decided to work together, mindful of each other's jobs.

What's the Deal With Solar Panels?

It is necessary to reduce silicone solar panels to the nuclear level to understand how they produce electricity. Silicon has an atomic number of 14, which means that its nucleus has 14 protons and 14 electrons. Three circles pass around the middle using the traditional concept of atomic circles. There are two electrons in the innermost band, and eight are in the center circle. The outermost circle is therefore half complete, holding four electrons. This means that it will still receive support from surrounding atoms to fill up. They form a crystalline structure when they are related.

There is little space for an electrical current to travel with all of those electrons going out and connecting. That is why the silicone used in solar panels, combined with an ingredient such as phosphorous, is impure. Five electrons are accessible in the outermost phosphorus band. The fifth electron is considered a 'free rider,' capable of travel without a lot of encouragement. The number of free carrier researchers is increased by adding impurities in a doping process. The result is the so-called silicone type N.

Silicium-type N is what is on the solar panel top. The mirror below is opposite — silicium P-type. Whereas N-type silicone has one extra electron, P-type contains impurities from materials with one less electron, such as gallium or boron. That creates an additional imbalance, and when the P-type is hit by sunlight, the electrons begin moving to fill the voids. A balancing act which repeats itself and generates electricity again and again.

What Makes Up a Solar Panel?

Solar cells consist of wafers of silicon. They are a hardened crystalline solid made of silicone, which is the second most abundant ingredient after oxygen in the Earth's crust. If you're at the beach and see bright black spots in the sand, it's silicon. This spontaneously converts sunlight into electricity, as Ochs learned.

Silicium can be mined like other crystals. Scientists such as those at Bell Labs, manufacture silicone as a single uniform crystal in a tube, unroll the tube and break it into the so-called wafers.

Vikram Aggarwal, founder and CEO of EnergySage, a comparative shopping platform for solar panels, says, 'Visualize a circle.' The stick is sliced like a 'pepperoni, a sandwiched slice of salami, and very finely cut,' he says. Historically, that's when it was challenging — just too thick, waste, or too thin to be capable and vulnerable to breakages.

Such cells are made from monocrystalline silicon to make them as thin as possible to get the highest potential value out of their crystal.

The early expectations at Bell Labs were that the solar cells would be ideal for the next space race, says Margolis, so the premium to retain weight is locked down. The photovoltaic cells, they were called, are mounted in a lightweight encapsulation. The solar cells have been placed in a small encapsulate.

Only four years after the first functioning Solar Cell, on 17 March

1958, Vanguard 1, the world's first solar-powered satellite, was developed and launched by the Naval Research Laboratory.

Solar Panels Today

Photovoltaic cells are now quickly produced and cut by lasers with greater accuracy than any physicist at Bell Labs might have expected. While in the universe, they have found much more purpose and value on Earth. Therefore, instead of relying on weight, solar manufacturers have put a premium on strength and durability.

Efficiency is one of the most important focuses on any solar manufacturer — how much of the sunlight on each mile of the solar panel can be turned into electricity. Aggarwal claims that it is the 'basic math problem' that lies at the heart of every solar production.

'Say, on the roof of your hypothesis, you can have 100 square feet,' he says, 'when you are 10% efficient, that's less than 20% in this limited space. Efficiency means how many electrons they can generate silicone wafers per square inch. The healthier they are, the more competitive they can deliver.'

In 2019 solar efficiency increased to 20%. There is a clear upward trend, but one which suggests that Margolis has a silicon cap. Approximately 10 years ago, the solar energy output ran by roughly 13%.

Due to the natural existence of silicone as an ingredient, the upper limit of the solar panels is 29%.

The Future of Solar

Many physicists are working on a new material called perovskite, which Aggarwal terms 'truly thrilling.' The mineral, which was initially discovered by the Urals in western Russia, has lifted perovskite's eyebrows in research, ranging from a 10% output in 2012 to 20% in 2014.

But Aggarwal and Margolis both warn that it is still in its earliest stages. 'Laboratory performance has improved quickly, but the lab varies from the real world,' says Margolis. Much progress has been demonstrated by perovskite in clean environments, but rapid reductions in elements such as water have been shown, which it could experience in everyday use.

Margolis and his colleagues are designing a strategy that he terms 'solar plus,' instead of new materials. As solar energy consumption grows, it is improving how 'solar communicates with other buildings.'

Think about it in the city's brutally hot summer. You go to work at a desk, then return home in the evening. The air conditioner is hot and moist, and everyone else in town has the same experience meaning that the electric grid gets tight.

However, Margolis believes that solar energy could be saved and used to reduce the strain. 'When the sun is still running, AC will pre-run two hours before your return home and make your house cool.'

The same goes for a cold winter that runs the risk of frozen pipes. 'You can heat your water over the warm day and always clean up the dishes by using hot water or have a shower the next morning ... we're just starting to think about how you can integrate the solar system into it.'

Margolis is optimistic, despite challenges faced with solar control, such as gas pricing and the political climate favoring fossil fuels.

'We're at this point when the utilities and developers know that solar is getting high enough for us to cope with it.'

CHEAP SOLAR ENERGY FOR HOMES

In the supply of solar power to homes, connected to the local power grid has always been accepted as a practice. Can you imagine using solar panels to harness free and green energy from the sun? It can save you up to 80% of your monthly energy bills!

Now homeowners can use solar power for homes with new knowledge and technology breakthroughs. The best thing is that you can build your solar panels, and you can save money straight away.

There is an array of online manuals for the manufacturing of solar panels for your home. I would say that you should choose one most appropriate for your level of awareness and what you can afford. Digital manuals for DIY neophytes are available at prices as low as $50, with complete descriptions and video content also available.

Videos of Do-It-Yourself support novices and guide you through the process. Quality and the comprehensive package will also help you to demonstrate how you can successfully check your solar panels. You will be told about how solar cells are made, as well as how the solar cells are attached to full solar power in homes and how they can be built at home, just like the professional ones are.

Keep in mind that manuals can be miswritten and may bring you more frustration than assistance at times. A reputable manual contains information on how to contact technical support, which could be helpful.

If you choose to do it yourself, solar power for the house will be reasonably expensive, so you will have to budget for it and be prepared to put in the extra effort. You just have to follow the instructions in the manual to be successful. Make sure you first read the entire manual to see how it all works. Once you're ready to continue, visit your shop and buy the necessary pieces.

If you're a novice, you'd be better to set up your solar power system just for one weekend to test it out. You should tackle any issues that occur during that weekend and make it a positive, engaging experience for your entire family. The best way to save money on that mischievous monthly bill is to use solar electricity. There is no excuse not to take advantage of this DIY challenge, which helps you to take full advantage of solar power throughout your life.

MAKE SOLAR POWER FOR HOMES-SHOW ME HOW TO BUILD HOME SOLAR POWER.

Want to know how to easily make solar power for homes? Solar power for the house is now an efficient and inexpensive way to produce electricity, mainly because there is plenty of sunlight to take advantage of. It's an approach well known for its efficiency and the provision of free power.

1. How Much Free Electricity Can You Expect to Get From Building Homemade Solar Panels?

The amount and scale of the electricity generated depends entirely on the environment and the way your solar panels were built. Depending on the quantity and form of energy used, the heat generated by residential solar panels can be used for many purposes.

2. How to Build Residential Homes Solar Energy?

Learning how to build a solar power system can be frustrating for beginners, particularly when trying to figure out the system's scale. I downloaded a high-quality internet guide, which taught me how to do this, whilst also helping me to learn about the principal components to create a photovoltaic process that generates electricity free of charge. The guide also showed me several tricks, such as isolating my house to reduce my energy consumption, which lowered my fuel costs.

3. How Does Solar Power Work for the Home?

To transform sunlight energy into accessible domestic electricity, photovoltaic cell technology is required. Within such photovoltaic (PV) elements, solar panels are installed with convert photons, as sunlight reaches them it creates direct current electricity. To generate an electrical current, solar panel cells are assembled and attached in sequence.

You may be able to see solar panels in sequence on the rooftops. The number and scale of the panels you choose will depend on the overall electrical energy you want to produce.

HOW DO YOU CREATE YOUR OWN, HOME-BASED SOLAR POWER GENERATOR?

Are bills for your power a threat? Is your spending a burden? Then the home solar power system is the ultimate solution. This research provides you with the power to produce for your house, but it also reduces your carbon footprint by using today's cleanest and most advanced renewable energy sources.

One of the reasons people used to avoid it was the high cost of building the solar energy system. But that's only the case if you are using professional companies to assist you in building a device as it could take up to a decade to get your investment back.

But you can build your own solar electric power generator more efficiently and conveniently. Some websites will guide you through every step of the process and you can own a solar generator by paying a fraction of the cost.

You may think the work for an unqualified person is tedious and challenging, but I'm confident that the site's guidelines will help you in the process. The manual is straightforward to follow step by step. Every piece used is easy to find in a hardware shop or the DIY store. To do this work, you don't have to be technically astute. It is straightforward by building a home-made solar power system that meets the specifications and requirements.

It will demonstrate how to effectively reduce the cost of high electricity bills by using natural and renewable energy sources. Get a cheaper natural electricity system, reduce electricity dependency, and turn it into green energy solutions.

SOLAR POWER SYSTEMS RESIDENTIAL-CREATE YOUR SOLAR PANELS AND CREATE SOLAR ENERGY

Although many breakthroughs have occurred in the use of residential solar power, very few homeowners jumped into solar-powered houses straight away. The main reason is that commercially constructed systems have high prices. A pro will keep approximately $20,000 for the rooftop mount when you hire the professional to build or upgrade a device. Once again, few are aware of the benefits of DIY solar power.

Learning how to install your solar panels has many advantages in life. You are released from heavy monthly utility charges. Besides living off the grid, the utilities energy credits for the excess energy you generate and the sale of excess energy back to them can actually earn you money. And for those who take their jobs with carbon tax credits, the Obama administration has the green package. Your home has a much higher value in time in-home assessments, which means that a full solar panel for your home system can be easily distributed for much more than your neighbor.

A good instruction manual is key to building your solar modules. DIY solar power information is readily available on the internet, but it is not very easy to find guides containing enhanced information. An excellent handbook gives you comprehensive, concise, and easy-to-follow diagrams with instructions. A collection of images should be beneficial when learning visually. You can follow the videos step by step, from the assembly, installation and maintenance. A genuine kit will also include community user's forum, spare time, and web service manual upgrades. Look hard enough, and you'll find one that suits these needs. Honestly, it's not too hard to handle these packages. And the beauty is that in your friendly neighborhood hardware store, you can easily find the items required for around $200.

Make the best choice that you could ever get and leave the biggest ecological legacies behind whilst spending $200 to get a lifetime of free energy. There is no denying that many people are going after the solar house, and by taking the lead to use DIY solar electricity, you have absolutely nothing to lose. Be one of the first ones to live well off the grid.

HOW TO INSTALL SOLAR POWER SYSTEM - WHAT YOU NEED TO INSTALL SOLAR POWER SYSTEM IN YOUR HOME

Wondering how to install and don't know where to operate a solar power system? In practical terms, here in this short essay, I would not be able to explain everything you need to know about installing solar electricity. Nevertheless, I will give you an overview of the topics you must commit to, particularly when you are thinking of a DIY project.

Before we begin, the mentioned topics can be found on the internet, but, you may be overwhelmed by the sheer amount of information you get before you even start because of the tons of information there. False information from outdated sources can also be found. This is much more effective from my perspective to look with a reference that has not only all the essential topics but has proven documents. To DIY novices, it must always be easy to understand and obey. Let's see what information you'll need:

1. First, a description of how the program works and alternative approaches to implement the system depending on specifications and the environment, if appropriate. It is essential to continue with the system's theories. There shouldn't be any time spent here. It would be

too tough for you as a novice because you'll take too long to absorb the entire program.

2. How to grab the pieces. You may think it's expensive to develop a solar power generation network. In reality, you can easily construct the whole device using components in your local hardware store. An excellent guide should inform you how to mount the best components on the lowest price solar energy network.

3. Installation of the solar energy system. This includes specific instructions on how to attach the parts, with electrical specifications and configurations and troubleshooting in mind. A complete novice can learn how to install a solar energy system, and it doesn't take months. The decisive factor is whether you are prepared to spend a couple of hours a week on a permanent electricity supply project.

4. You need to monitor how much energy is generated by the system and how much energy is stored to use it later. Without these two components, the system is complete.

Other valuable subjects also include installing on-grid and off-grid solar power systems as well as wind turbine power generation.

DIY IS THE MOST COST-EFFECTIVE WAY TO BUILD SOLAR POWER HOUSE

You've got to know solar energy. Often when you watch movies, you hear this word. Sometimes when you read newspapers, you see this word. You'll probably have seen solar powerhouses near you. You can see solar panels on your neighbor's roof. It is a trend, as solar energy firms can bring us many advantages. You can save money by cutting electricity bills every month and even protect our environment because solar energy is an electricity source that is renewable and clean. You'll learn more about solar energy houses in the next few minutes.

It certainly has to do with a lower building cost for each homeowner to build its own solar-powered house. In the past, you wouldn't like to build a solar house because it was too costly. Even if you wanted to be environmentally friendly, you didn't want to invest. The rate is now slightly lower though. A solar energy network costs just $3000 for households to install. You can spend the money on recruiting professionals for the installation of the system. Naturally, it's only one way but most people want to spend less.

You can do it yourself. Perhaps, in the past, you wouldn't know about creating your solar power system. There are several guides on the internet for DIY solar energy houses for you. Experts of home solar systems produce them and show you how to comfortably construct a solar functional device for less than $200 by adopting their hidden methods. In reality, you have to spend more time and energy because you pay less money. Usually, you will spend a whole day watching the videos of the step-by-step process and building the programs.

A home solar power system should power all your electrical equipment such as your microwave, refrigerator or TV. Most find that with a solar powerhouse, they can reduce at least 50% of electricity bills. I strongly recommend that you build your system on your own if you want to save money too. I'm sure that you can do this with the help of an advanced DIY solar powerhouse guide. You can begin thinking about how to use your extra money once your home-made solar energy system is ready.

DIY SOLAR POWER FOR HOMES

I'm sure you've recognized the great resource for our planet, that is solar energy. But, did you know that it's good for you and your family to have solar power for a home? More and more consumers are now realizing this and are trying to install solar panels for residential purposes.

5 Reasons Why You Should Love DIY Solar Power for Homes

1 – It is straightforward now to know how to install a home solar power system. To learn how to install the systems yourself in a few days and at lower costs (for less than $200!) and to buy DIY solar electricity company manuals for your projects!

2 – In 2009, the US government decided to offer energy efficiency Federal Tax Credits *. The tax credit for households installing solar power systems would cover 30%, with no limit to 2016 (both existing homes and new buildings), thus lowering costs for building the solar energy system.

3-Solar conversion efficiency is rising, and prices are decreasing thanks to the development of solar technology. This means you can pay less money plus produce better household energy.

4 – America is in the world's top five nations that embrace domestic solar energy. Solar energy is currently installed in only 20,000 households. (Reuters, 2009) So you'll be the pioneer of your peers in

the use of renewable energy. Your effort to protect our world will be appreciated.

5 – People are now aware that oil and coal are the primary sources of power, causing severe pollution and other environmental issues. We are finding alternative energy to save the world.

Save and Earn Money with a Solar Power System for Home

Does home solar power sound attractive? The main advantage of household solar power is that it saves you tons of money every month. On the electric bills, you will generally save more than 80%.

Moreover, you're going to earn more money. This is because you can sell extra power to the electrical supplier if you produce more energy than your home has used. The organization will buy additional energy from you according to regulation.

Install A Solar Power System for Under $200

It can now be very cost-efficient to produce solar energy systems for your home. You can contact other corporations if you are wealthy but what if you have a tight budget? In reality, you can create your own program.

You can note that the cost of solar panels can be reduced from $1000 to $200 with plenty of solar power initiatives. That is why many people think about the creation of their home-grown solar system.

Power4Home is currently the most popular service. You can create a robust power system for less than $200 in this software.

Enjoy Tax Credits & Lower Building Costs

You can also benefit from using solar energy by breaking off when solar power is installed for households at tax time. The US administration is aimed at promoting 'green energy' and environmental conservation.

If you have a solar power plant, that means that you will have a reliable, safe, and permanent energy source in the future.

You don't have to spend much money on the solar power system. The need to remove or replace the pieces is rarely required and that saves you a lot more money in exchange.

Make A Cleaner Environment for Our Children

To think big, with home solar power, allows you to be one of the people who will protect your planet for the next generation. You won't create air pollution anymore if you use renewable technology.

Solar energy usage will keep the world from being more polluted by using environmentally sustainable resources. Your children will appreciate what you choose to do today.

Solar Power is The Best Solution

We have also found that the energy supplies that we use today will very quickly vanish, and the only alternative is solar energy. Not just recognizing the crisis, many people take measures to get solar electricity to their homes.

The disaster that you are avoiding by using renewable energy resources is very easy to imagine. So why don't you begin planning?

These are only the essential benefits of household solar power. When you spend time on in-depth research, you will find many other advantages over using solar energy. This will increase your drive to create your home with a solar energy network, and you will know that renewable energy is our potential energy supply.

SOLAR HEATING

Active solar heating - a dynamic solar heating systems heat a fluid with solar energy, whether liquid or air, and the solar heat transfer directly into the internal space or into a corresponding storage system. When the solar system cannot supply adequate heating, additional heat is provided by an auxiliary or back-up system. Liquid systems are used more commonly when transported and are ideal for radiant heating systems, hot-water boilers, including immersion heat pumps and cold storage systems. Air and liquid systems can supplement air systems that are forced to operate in air.

Liquid-Based Active Solar Heating

Central heating is ideally suited to liquid solar collectors. They are the same as solar household water heating systems. The most common platform collectors have is evacuated tubes and concentrating collectors. In the reservoir, solar heat is absorbed by the transfer of heat or 'used' fluids, such as gas, antifreeze, or other fluid. A controller operates a pump to move the liquid via the collector at the appropriate time.

The liquid fluid flows quickly, and its temperature rises by the collector to just 10 ° F at 20 ° F (5.6 ° to 11 ° C). The thermal loss of the collector increases by heating smaller amounts of liquid to higher temperatures and reduces system efficiency. The liquid flows into a storage tank or an external heat exchanger. The piping, pumps, valves, an expansion tank, a heat exchanger, a storage tank, and controls also belong to the system.

The rate of flow depends on the fluid of heat transfer. See solar water heating for more detail on fluid solar panels types, size, maintenance, and other issues.

Storing Heat in Liquid Systems

Liquid systems store solar heat either in water tanks or in a radiant plate system's masonry mass. In the storage tank type system, heat transfers from the working fluid to a delivery fluid in or outside a heat exchanger.

Depending on the overall system design, the tanks are pressurized or unpressurized. Consider cost, scale, reliability, location (within or outside the basement), and the building of the storage tank. If a tank of the required size doesn't fit into existing doors, you may have to build a tank on site. Tanks must also meet local houses, plumbing and hydraulic standards with restrictions on temperature and pressure. You also need to be aware of the insulation needed to avoid unnecessary heat loss and the type of protection or filtering necessary to prevent corrosion or leakages.

In systems with general storage requirements, modified or tailored tanks can be required. They typically consist of stainless steel, fiberglass, or plastic at high temperatures. The tanks are also available for concrete and wood (hot tub). Each type of tank's advantages and disadvantages and the size and weight of each type require careful placement. It may be more practical to use several smaller tanks rather than one large one. Standard domestic water heaters are the simplest storage system option. They comply with pressure vessel building codes, are lined for corrosion inhibition, and can be easily installed.

Distributing Heat for Liquid Systems

The solar heat can be distributed via a radiant floor, baseboards, or hot water radiators or a central forced-air system. The solar-heated fluid circulates in a radiant floor system via piping in the floor, then radiating heat into the room. Floor radiation is ideal for liquid solar power systems because its performance is relatively low. A carefully built system may not have a separate heat storage tank, but most devices can operate it. A conventional boiler or a modern home water heater is capable of supplying heat backup. The layer usually is tile completed. Radiant plate systems take longer than other heat systems to heat the home from a 'cold start.' However, once operated, a consistent heat level is provided. The efficiency of the system would be decreased by taping and rugs. For more detail, see Radiant Heating.

Warmwater foundations and radiators are water-required for effective heating between 160 ° and 180 ° F (71 ° and 82 ° C). The flats are usually heated between 90 ° and 120 ° F (32 ° and 49 ° C) for moving and handling fluids. Therefore, using solar heating system baseboards or radiators allows the area of the baseboard or the radiators to be wider, a contingency system to increase the solar-heated liquid temperature, or to cover a solar-heated collector with an evacuated tube collector.

A fluid pump can be built with a forced-air heating system in many ways. The basic concept is for the central room-air return duct to place a liquid-to-air heat exchanger or heating wire until it enters the stove. Air coming back from the living rooms is heated by the solar fluid in the heat exchanger when it goes in. The furnace provides additional heat, as needed. The spindle must be large enough that adequate heat is transferred to air at the collector's lowest working temperature.

Ventilation Preheating

Solar air heating systems use air to absorb and transmit solar power as a working fluid. Solar air collectors may heat individual rooms directly or probably pre-heat the air from an air source thermal heat pump or an air coil.

Air collectors capture heat early and later than liquid systems and can provide more usable energy than fluid systems of the same scale over a heating season. Besides, air systems do not freeze, as opposed to liquid systems, and minor leaks in collectors or distribution ducts do not lead to serious problems, although performance degradation can occur. The air is less effective than the liquid heat transfer medium, so solar panels are less efficient than liquid solar panels.

Early systems move solar-heated air through the beds of rocks as an energy storage system. This solution is not recommended due to their lack of capacity and the possible problems of condensation and mold in the rock bed and the impact of moisture and mold on indoor air quality.

The solar air collectors often have their presence concealed in the walls or roofs. For example, an air-flow system could be installed into the tile roof to use the heat absorbed by tiles.

Most solar air heaters are room air heaters, but there are small applications for new equipment called sweat air collectors.

Room Air Heaters

For heating one or more rooms, air collectors can be installed on a roof or an outside (south-facing) wall. While on-site factory-built collectors are available and do-it-yourselfers may opt for a separate air collector. It will cost a few hundred dollars to produce a single-window air heater.

There is a black metal plate for the absorption of heat in front of the collector with an airtight, glazed metal frame. Sunlight heats the flat and, in effect, heats the collector's air. An electric fan sucks in the air from the room and blows it into the collector's door. Detached collectors require conduits between the room and the collector to carry air. Wall-mounted collectors are placed directly on an adjacent wall, and the air inlet and outlet panels are cut through the wall.

Simple window box collectors fit the window opening in an existing window. We can be active or passive (with a fan). Using passive methods, the air enters the collector's surface, rises when heated, and goes into the house. The room air does not flow back into the panel via a baffle or damping (reverse thermosiphon) when the sun is not shining. These systems provide only a small amount of heat, as the collector area is quite low.

Transpired Air Collectors

Sweat pickers use the necessary technology to capture the heat of the sun into hot homes. The collectors are made of dark metal panels that are mounted on the south-facing wall of a building. Between the old wall and the new façade, an air space is created. The dark outside façade absorbs solar energy and heats up quickly during sunny days — even in the cold outdoors.

The ventilating air is pulled into the building by a fan or blower through small holes in the collectors and by the air between the collectors and the south walls. The solar energy absorbed by the collectors heats the airflow to 40 ° F. In contrast to other heating systems, transpired glazing systems do not require costly glazing.

Transpired air collectors are suitable for large buildings with heavy ventilation loads, which typically don't fit tightly sealed new homes today as there's less airflow from the outside. However, it is possible to use small air collectors to pre-heat air into a heat recovery ventilator or to heat the air coil to the heat pump of an air source, thus improving its efficiency and convenience on cold days. However, no knowledge about the cost efficiency of using a contemplated air collector in this manner is currently available.

Efficiency and other advantages of active solar heating systems

Strong solar heating systems are more cost-effective as they shift the more expensive heating fuels, such as power, propane, and oil, in cold weather with sufficient solar energy. Some countries provide exemptions from sales taxes, income tax credits or deductions, and tax exemptions or deductions on the property for solar energy.

The cost of an active solar power system can vary. Commercially available dealers have ten years warranties, and a good solar power system can last comfortably for decades. When domestic water is heated, the economy of an active space heating system improves because, in summer, an otherwise idle collector can heat water.

If you heat your home with an active solar system, your fuel consumption in winter will be drastically reduced. A solar heating system often decreases air emissions and greenhouse gases arising from the use of fossil fuels for heating or power generation.

Selecting and Sizing a Solar Heating System

The selection of the appropriate solar energy system depends on factors such as size, design, and heating needs. Community arrangements may restrict the options; for example, councils may not permit solar collectors to be built in certain areas of your house (though many homeowners have successfully challenged these conventions).

The local atmosphere, collector size, capacity and collector area decide how much heat can be provided by a solar heating system. The installation of an active system to provide 40-80% of house heating needs is typically the most economical approach. Except for solar heater collectors that heat one or two rooms and do not require a heat store, systems that provide less than 40% of house heater are rarely economical. A well-designed and isolated building with passive solar heating techniques will demand a smaller and less expensive heating system. It will require very little additional heat other than solar methods.

In addition to the fact that it is typically not feasible or economical to design an active system that supplies enough heat 100% of the time, the majority of building codes and loans require a backup heating system. Additional or alternative power sources where the energy needs of the solar system cannot be complied with require backups that vary between a wood stove and a typical central system of heating.

Solar Heating Systems Controls

Solar heating system controls are generally more complex than traditional heating system controls because more signals and more equipment (including the central backup heater) have to be analyzed and controlled. Solar controllers include system operation sensors, switches, and/or engines. The system uses other sensors to prevent freezing or very high collector temperatures.

A differential thermostat tests the temperature difference between collectors, and the storage system at the center of the control system. The thermostat switches on a pump or fan to move water or air via the pilot to heat the storage medium or the house when the collectors are 10 °F to 20 °F colder than the storeroom.

These controls vary in operation, performance, and cost. Some control systems monitor the temperature in various parts of the system to assess how it works. Microprocessors monitor and automate heat distribution, and transmission in storage and the house areas are used for the most sophisticated devices.

A low voltage, direct current (DC) blowers (for air gatherers), or pumps (for liquid collectors) can be used with a solar panel. The output of the solar panels corresponds to the solar collector's solar heat gain. The blower or pump speed is designed for efficient solar production in the working fluid with proper sizing. A blower or pump speed is sluggish under low sun conditions and runs faster during high solar gains.

Separate controls may not be necessary when used with a room air collector. In case of a power outage, this also ensures that the system is running. A battery-storage solar power system can power the central heating system, even if it is expensive for big systems.

Building Codes, Covenants, and Regulations for Solar Heating Systems

You should look at the local building codes, zoning regulations, and subdivision pacts, as well as any special regulations relating to the site before installing a solar energy system. A building permit is probably needed to install an existing building with a solar energy system.

Initially, not every community or town welcomes renewable residential energy plants. Although it is often due to ignorance or the comparative novelty of renewables, you need to follow existing buildings regulations to allow your system to be installed.

The topic of building codes and zoning for an implementation of a solar panel is typically a local concern. Your city, province, or parish usually enforce this, even if a statewide building code is in effect. The common problems with building codes for homeowners include:

- Exceeding roof load
- Unacceptable heat exchangers
- Improper wiring
- Unlawful tampering with potable water supplies

Potential zoning issues include these:

- Obstructing side yards
- Erecting unlawful protrusions on roofs
- Siting the system too close to streets or lot boundaries

Also, compliance is required in the special field regulations — such as the covenants of the local community, subdivision, or homeowner association. The agreements, historic district regulations, and provisions of the floodplain can easily be ignored. Contact the zoning, building enforcement, and any appropriate house owner, subdivision, neighborhood, and/or community association(s) within your local jurisdiction to determine what is needed for local compliance.

Installing and Maintaining Your Solar Heating System

It depends on the effective positioning, system design, installation, and the quality and durability of the components that an active solar energy system performs. Collectors and controls are of good quality today, but having a qualified contractor who can correctly build and configure the machine can also be a challenge.

Once a system has been implemented, its performance must be appropriately maintained, and faults prevented. Different installations require specific maintenance forms, and you can set up a calendar that lists the maintenance activities recommended for your installation by component suppliers and installers.

Many solar water heaters are protected directly by the insurance policies of the homeowner. But, usually, no damage is caused by freezing. You'll also need to find out what your coverage is, call the insurance agent. Even if your supplier protects the program, it is better to tell them in writing that you have a new system.

SOLAR COOLING

Solar cooling is a device that transforms sunlight heat into a cooling system that is suitable for cooling and air conditioning. A solar refrigeration system collects and utilizes solar power in a thermally driven refrigeration process that, in effect, lowers and regulates the temperature for applications such as producing refrigerated water or cooling air for a house.

There are several different techniques for cooling cycles with different principles. Three techniques are among the most popular:

- Absorption cycles
- Desiccant cycles
- Solar mechanical cycles

How Solar Cooling Works

No matter the strategy used, three main elements are usually used in the solar cooling system:

- A solar collector for heat or mechanical work, like a solar panel, used for converting solar radiation.
- A cooling or air conditioning facility used for cooling production.
- A heat sink that absorbs and irradiates any heat that has been discarded.

How Solar Cooling Works

The ultimate objective remains the same as the techniques used to achieve solar cooling. It uses an external heat source such as a solar panel to collect the atmospheric temperature and then uses the heat with a coolant to produce pressure in a closed cooler circuit, enabling a solar cooling system to operate.

Refrigerants are fluid or mixture that removes heat from the environment and, if combined with other elements, including compressors and evaporators, may create a refrigeration or air conditioning system. In most cooling cycles, the coolant moves from the liquid to the gas phase and reverts to its cooling aims.

The cooling process is based on the evaporative cooling of a refrigerant during absorption cycles. The process takes heat from the system, leaving the remainder cooler than before as the vaporization requires energy input. Absorption cycles complete the pressure process by dissolving the refrigerant from a hydraulic pump, or anything that quickly consumes air.

Four primary elements include an absorber, compressor, condenser, and evaporator to aid the absorption cooling process. The evaporator is mainly the cooling or air conditioning system since it is used in all cooling systems.

The cooling process progresses as follows during an absorption cycle:

The absorber contains a mixture of absorbents and coolants, which is delivered via a liquid pump into the generator.

The generator uses and heats the absorbent-coolant mixture with external solar energy collected from a source such as a solar panel. As a response to the heat, the solution starts to boil and converts water into vapor.

For processes obtained by the heatsink, the condenser liquefies the water vapor and avoids the liquid. An expansion valve channels the fresh liquid condensate into the evaporator.

Eventually, low-pressure evaporation of the coolant contributes to the heat being consumed in the chilled room and to the cooling effect.

By the conclusion of the process, the vaporized coolant returns to the absorber. This cycle is powered by solar power.

Desiccant cooling systems rely on cycling processes for dehumidification and moisture removal. It uses materials and substances that dehumidify water from the environment. Such products are referred to as dryers. The dryers are regenerated by the use of solar power in the process.

Desiccant refrigerants may work with solid and liquid dryers. The process of drying cooling is followed by:

1. The desiccant absorbs the water vapor, and the air in the process is dehumidified or absorbed by the unit. A transition occurs from

vapor pressure separation, which releases heat due to water condensation, which induces a heat exchange.

2. In the vacuum or in an evaporative cooler to refresh more, the air is then moved into the regenerator by the filtered desiccant. However, it has to pass a liquid-liquid heat exchanger and a heating roll to increase its temperature before the diluted desiccant can enter the rectifier.

3. The regenerator must be exposed to regenerative air in the hot, distilled desiccant that allows moisture to move from the condensed solution into the liquid. The resulting difference in steam pressure is this transfer.

4. The resulting, more concentrated dryer goes back through the liquid-liquid heat exchanger, and the cooling belt then returns into the dehumidifier, which repeats the cycle.

Solar mechanical cycles, the third technique, function differently from absorption and drying cycles. The goal of solar mechanical cycles is to merge solar energy dynamics with traditional cooling systems, instead of creating an entirely new system. In this cycle, solar energy is used to power the actual motor, which produces energy for the cooling system as a whole, as it does for the absorption and drainage cycles, instead of flowing into the absorption chiller.

Applications of solar cooling

The primary objectives of solar cooling are: to cool food storage or cool space in general, or for air conditioning. In cars such as RVs and campers that use the cooling system, solar cooling may be observed. The use of solar cooling is also demonstrated by vapor absorption cooling systems used in industries where shallow process temperatures and high thermal capacity is needed.

Perhaps the biggest use for solar cooling is its potential to provide cooling systems for countries that would otherwise be unable to deal with the high electricity and energy costs and strain of traditional cooling systems. Solar cooling decreases significantly, the amount of energy necessary for refrigeration uses, such as vaccinations and agricultural products, thus saving costs and improving the atmosphere by using green energy and reducing the use of materials for ozone depletion.

Challenges of solar cooling

Domestic cooling systems often are not economical, whereas solar cooling is used in different industrial environments. High cost and low performance from internal systems were a significant hurdle on their broader domestic use.

While the system's long-term operating costs are less than those of traditional cooling systems, initial investment costs are much higher as system components like the solar collector and storage tanks are supplied in limited quantities at cheaper rates.

SOLAR PANELS TO POWER THE GROWING GLOBAL DEMAND FOR AIR-CONDITIONERS

A new report shows that electricity demand in the emerging economies to feed the rapidly growing use of air conditioners will nearly double by 2050.

New Energy Outlook 2019 says Bloomberg New Energy Finance (BNEF) estimates that by the middle of the century, air conditioning will make up 12.7% of the world's demand for electricity.

The report also shows, however, that solar panels can provide considerable demand for this addition. The use of batteries also facilitates cost-efficient night-time air conditioning.

Why Solar Installations Are Perfect For Air Conditioners

BNEF also states that by the mid-century, wind and solar power will account for 50% of global electricity generation. The relative wind-power mix over the next 30 years is presented in a chart.

Renewable growth helps offset the expected 93% rise in energy demand in emerging economies for air conditioners. This is due to population growth, higher revenues, and reduced cost of equipment.

But, when it comes to air conditioners, solar panels have a particular benefit. This is because the heat of the day when the sun shines brightly needs optimum cooling.

This will lead to different peaks for countries with the highest demand for climate change. These move from cooler afternoons to mid-après midnight when solar systems produce optimal power.

Battery Storage Is Rapidly Growing Worldwide.

Battery storage will also change energy markets as we hit the century, according to BNEF.

- Since 2010, battery costs have plummeted 84% already.
- Constant price reductions are the result of battery production for electric vehicles.
- The most cost-effective method of peak generation will be batteries before the mid-2020s.

Batteries would replace, coal, and gas in dispatchable production by 2030.

- More residential solar batteries will be purchased from 2025.
- In the next 20 years, payback periods will halve for solar battery Systems.

BNEF expects a five-fold increase in the demand for small-scale batteries. This will increase the market from $2.1 billion in 2050 to $10.7 billion.

BNEF also notes that consumers are gradually using solar batteries

– including Tesla Powerwall 2 and Enphase – as interactive power stations.

When the Sun does not shine, and direct solar energy isn't possible, batteries facilitate cost-effective use of air conditions in the evening.

Solar power is winning the cost argument with coal-fired energy

By 2030, prevailing wind and solar projects will be less costly than almost everywhere, running existing oil and gas plants. As soon as 2027, China will hit this coal tipping point.

The report notes that the power system of Australia is rapidly becoming the world's most centralized. That's why domestic solar panels and batteries behind the meter will make up 38% of all power capacity by 2050.

Nearly all of the coal-fired generators from Australia will be retired by then, BNEF says. This will lead to a decrease of around 83% of harmful emissions.

INVERTER BASICS

What Are Inverters?

An inverter is a device that translates battery power (DC) into a higher voltage AC. This means that the majority of inverters are installed and used together with a specific sort of battery bank – a typical setup in off-grid solar facilities.

The heart of an off-grid inverter powered electrical system is deep cycle batteries that can store energy on demand. Direct current (DC) at the nominal voltage of the battery is the most common method of producing electric power. For example, your car radio uses 12 volts DC, the same voltage as the car's battery.

Many off-grid power plants (not operated by a power supplier) use 12-volt DC power to carry out essential loads, such as lighting. These systems are commonly referred to as low-voltage DC systems. (The electrical power consumption is called a load). A 12-volt DC network allows you to take advantage of electrical lighting, gaming facilities, desktop computers, and other gadgets that can be powered by a vehicle battery.

However, without the help of some device that generates 'household' electricity, or AC power, you cannot operate power tools, kitchen appliances, or bureau machines. The norm is 120 volts of current alternating in North America (120 V AC). 230 Volts AC

50Hz is the standard in many countries.

One option is to use a battery-driven generator to operate your primary machines. Fire it up, and you can use your devices and tools as long as it is running (noisily). But the electricity stops as soon as you disable it-or it runs out of fuel.

Another option is the production by inverter of 120/230 V AC from batteries. With an inverter, the battery power (DC) is converted to a higher voltage alternating current (AC). There have been DC-to-AC inverters for a long time. In the beginning, energy loss was very high; the average performance of the inverters was around 60%. It means that 100 watts of battery power would have to be drained to power a 60-watt lamp.

In the early 1980s, we developed a new method for constructing inverters. Such sufficiently stable inverters have improved performance to 90%. This technique was pioneered by trace engineering. Their first iteration, the DR1512, was launched in 1984, and in every corner of the world, there are already thousands of their firstborn.

The beauty of their architecture is the secret to Trace inverters' durability (Trace was subsequently bought by Xantrex, who was then bought by Schneider Electric). We use an innovative transformer effect field (FET) to convert the DC voltage of the battery (typically 12 or 24 V DC) in AC. And a low voltage AC, usually 120 or 220 volts AC, is converted into a higher voltage. All of the power formation-converting form the transformer's low voltage side to AC- and waving shaping.

One cautionary note is that power use is easy to relax when connected to an electric grid. The idea goes, the utility provider would certainly not go out of business as long as you can afford it. The sum of electric power is, however, limited for a battery. You will measure and control the electricity usage when you do not want to run the battery(s) down.

How Solar Inverter Works and Its Applications

Nowadays, renewable energy sources, such as solar have become even more critical. A solar inverter is a typical inverter, but it uses solar power from the Sun. This form of inverter helps to transform DC to AC with solar power. DC is the power that flows in one direction and helps to supply electricity when no electricity exists, for example; compact machines, such as smartphones, iPods, MP3 players, etc, with the power contained in the batteries that are directly powered. The power supply inside the circuit is the AC (alternative current) method. A solar inverter allows specific DC-driven systems to run on AC power such that the user uses the AC power. The AC power is typically used for home equipment. If you think why this inverter is used as an alternate to the usual electric converter, it is because the solar inverter uses the solar energy that is generously obtainable from the sun and is clean and pollution-free.

In a solar power system, the Solar Inverter is an integral tool. The fundamental function is to transform the direct current variable output of the solar panels into the alternating current—various electrical and electronic components aid conversion in the circuit.

The alternating current power conversion is used to control appliances like the TV, fridge, microwave, etc. We may correctly use the direct current electricity from the solar panel, such as a cell telephone adapter for specific particular applications. The capacity of a home solar panel is usually used to fuel AC electricity.

Types of Solar Inverters

There are several prominent solar inverter models in the world. In the industry, however, there is a range of solar inverters, which include the following.

Off-Grid Inverters

Off-grid inverters are used to supply the DC power from the battery panel with the solar inverter. These battery panels are charged via solar panels. Several solar inverter systems have been built into the system using a basic battery charger, capable of boosting battery power from the AC source.

Grid-Tie Inverters

An inverter that is associated with the grid can be said to be a grid-tie inverter. The inverter supplies electricity in the power grid by the matching frequency and phase. The rate of o / p AC power with the efficiency provided AC power range is 50Hz in India and 60Hz in North America. These wheelers have been developed to shut down automatically when a power supply loss is detected.

Battery Backup Inverters

These particular inverter types are designed especially for drawing battery energy. An on-board charger preserves the charge of the battery and transfers additional energy to the grid. These inverters can provide AC power to different loads during service interruptions. The anti-insulating function also applies.

Micro Inverters

In the solar industry, micro-inverters are modern. They are small, compact, and portable, performance-consistent. All features of any central inverter are included.

Solar Inverter and It's Working

This project's primary objective is to develop a solar energy system for domestic charges via an inverter. The most critical hardware specifications include Bridge MOSFETs, Step up Transformer, Voltage Controller, MOSFET Motor, PWM IC, Solar Panel, and Battery.

Using photovoltaic cells, solar power is converted to electrical energy. This energy is collected during the day in batteries for use as required. The suggested system is designed to use solar energy with an inverter for domestic charges.

A solar inverter can transform the DC (Direct present) output of a PV solar panel into an AC (alternate current) utility frequency, which can be supplied to a central, off-line electrical n / w industrial electrical grid (or to) system.

The solar energy is deposited in the photovoltaic battery of this proposed device. With the MOSFET driver inverter MOSFET, this battery energy is altered to AC's 50Hz frequency supply, and the voltage is recharged by a transformer, which is not grid binding but is all off-line NW.

For example, the SMF type and 5 AH battery (unsupplied, generally used in Small UPS) should be used as a high power casting solar cell requirement. Besides, this project can be added to the overvoltage, voltage safety, and overload protection load system.

Advantages of Solar Inverter

The different benefits of the solar inverter must be addressed after learning in detail what a solar inverter is and how it operates on the residential and industrial ground.

The greenhouse effect and global warming are continuously being reduced by solar energy.

- It would help to save money and even electricity through the use of solar-powered devices since others have come to use

these apps.
- The DC is converted to battery or AC with a solar inverter. This supports people to partially use electricity.
- The solar inverter synchronous, which helps small homeowners and power companies because they are massive.
- Solar multifunction very carefully converts DC to AC, which is ideal for businesses.
- The inverter, i.e. low costs generators, are economically efficient.
- Solarev energy can be used to power everyday devices in your home or workplace.

Solar Inverter Disadvantages

- Primarily, a lot of money needs to be spent to purchase a solar converter.
- It works well and only produces DC when the light of the day is high.
- The solar panels are only useful when exposed to large amounts of sunshine.
- Solar inverters will operate when it's sunny but the available battery is fully charged with sun's assistance.

DIY SOLAR POWER KITS - CHEAP & EASY WAY TO USE SOLAR ENERGY FOR HOMES

Do-it-yourself (DIY) solar kits are a perfect idea for your home if you want to work with your hands and perform tasks that are both beneficial for your budget and the world. They are becoming increasingly popular as more and more people try to cut the cost of living after a difficult global economy and reducing their environmental impact for future generations.

DIY solar power packages aren't hard to put together, despite what you might think. With an easy-to-follow guide, putting your package together and saving energy will begin by tapping your solar energy source.

Cheap and Easily Available

DIY solar kits are inexpensive, and the components are less than $300, and in some cases, as little as $200, are sold in a local hardware store. That means tremendous savings on solar market panels, which can cost several hundred dollars or even thousands.

Great Family Activity

Including the whole family in the process is another critical benefit of building up the home-made solar energy generator. You'll enjoy working as a team to bring the supplies together, and the entire family will be excited to see how the solar array works once it is installed. This is a great way to teach children how to reduce the consumption of vital energy and how to reduce environmental impact.

Reduce Home Energy Consumption

Many estimates indicate you can reduce the energy costs by up to 60% by installing solar panels in your house. However, many people believe that it is more reliable to install solar panels. This represents tremendous savings in any case, especially given the relatively low cost of DIY solar panels. These kits pay more in just a few months for energy savings.

Let's imagine, for example, that you pay $100 a month in electricity on average, depending on the climate and weather you have, your electricity expense would decrease to $60 a month if the energy bill was reduced by 40%. The home solar panel system will be paying for itself within one year.

Getting Started With Renewable Energy

There is no reason to delay beginning your own DIY solar power generation project for your house. Not only can you make it simple for Mother Earth by using green energies, but you can also help reduce your own carbon footprint. The materials are cheap and extensive resources are at your disposal to guide you in the assembly of the panels.

In the meantime, the benefits of the design of your home solar system would be apparent. Benefit from the profits of solar power to you, your family, and the world.

THE GUIDE TO THE INSTALLATION OF SOLAR DIY HOME

You might think that DIY home solar is too difficult for you to install.

Maybe you think it's too technical or too big a job for a novice. The reality is that almost everyone can mount a DIY solar pack.

Besides, a photovoltaic system is worth more than the long-term financial benefits and energy independence.

Why change to Solar Energy?

Homeowners must take into consideration the potential benefits of the construction of a solar panel.

The primary explanation for taking photovoltaic energy is the extraordinary cost savings.

You can save on energy costs throughout your system when your system is up and running. And for 30 years or more, you can expect your PV panels to work at or near full capacity!

Nevertheless, the gains will not end at savings alone. Inclusion of a solar panel array:

- Increases property values
- Enables freedom of energy
- Saves the environment
- Sets a good example for your friends and neighbors

So, you're ready to begin?

Where to Start DIY Solar

A solar DIY network needs to be very well prepared for. You need to ask these questions:

- Is there sufficient room for an assortment on your roof or anywhere else on your property?
- Where's the best place to put the collection for you?
- Will they be in range of the sun's rays throughout the day.
- Are there barriers, like sun-blocking trees or buildings creating shade?
- Can the weight of the panels be handled by the roof?

You are willing to continue, once you have answered these issues and have decided that solar power will work for you. And don't give up if you can't answer every question unequivocally 'yes.' There are many ways of working in the shade or even with a little roof if needed.

See Your Electrical Records.

Now collect your electrical bills over the last twelve months to figure out how much power you are using during your maximum month.

Look for the consumption figures of kWh (kilowatt-hours), and total for the year. Remember this number so you can see it later in your plan.

Your average monthly power usage determines the size of the device you need.

Choose your components for the DIY Solar Kit.

To build your system, you will need:

- Solar panels
- A mounting system
- An inverter

It should be noted that a battery safeguard is necessary for an off-grid system. This adds a degree of complexity to the DIY process, which is not recommended by most specialists unless you're experienced.

The DIY Solar Installation Process

Professional construction constitutes a significant part of the expense of a solar panel system, which saves you big bucks.

However, it is essential to bring a licensed electrician to the power plug. This is more costly than cording, but in most cases, it is not an alternative unless you are an electrician.

You have to plan the assembly and racking device first before the panels will be installed. These systems are plug-and-play systems, and they are quite easy to install, particularly using SnapNrack products.

When the racking is attached to the roof, the panels themselves are easy to assemble. You just need to put the panels in action, attach and bring them in place as you go.

Here, just wire in the frame, and that's it! You ought to be good to go!

Research reputable and well-established sources to buy your DIY solar kit. Only in this manner can your components be guaranteed in full.

Furthermore, a manufacturer of solar kits can be an invaluable tool to help you plan your device and answer questions through the whole DIY home solar installation cycle.

WHAT EQUIPMENT IS REQUIRED?

More homeowners around the globe opt for the installation of solar panel systems. They aim to lower energy costs in the long term while minimizing their carbon footprint. Clean solar energy is the future of energy, which is evidence of fast growth throughout North America, Europe, China, and even India. The energy is clean, renewable, and affordable.

An SEIA report states that in the three quarters of 2019, there was a record volume of residential solar power, with total growth estimated at 23% in 2019. In the coming years, growth is forecast to continue. This accelerated growth is primarily attributed to capacity upgrades and reduced prices of the new solar energy systems.

So, what is necessary to build an efficient home solar power system? What machinery is necessary? How are investments and savings supposed to be? When is break-even going to happen?

What Is A Solar Panel System?

Solar panel systems with roof-mounted racking absorb and convert natural-sunlight photons into a usable form of energy. PV or photovoltaic solar power modules are also called solar panels.

A high-quality solar power system can reduce or eliminate the dependence on the community electricity grid for lightning, warming, cooling, and running your house. It is a safe, renewable source of energy that needs minimal maintenance and will repay the initial investment in a few years.

A system with a 25-year warranty provides decades of free energy over the longer term.

What Are A Solar Panel System's Core Components?

The first step is to understand the elements of a solar system. In-house solar power or PV network components:

Solar panels are the main elements of a solar cell. Performance, quality, warranty, and technology are the main qualities to remember. 'SolarReviews' provides a comprehensive, biased list of leading global solar panel brands comparing the efficiency and guarantee attributes.

- Monocrystalline and polycrystalline are two types of solar panels that are ideally suited to residential solar systems. The monocrystalline works in a similar way, although it is less compact and a little more expensive.
- The number and position of solar panels depends on your needs for electricity, useable roofing surfaces, air conditioning and peak sunlight, the efficiency of solar panels, and the availability of a net measurement system. It usually also has the potential to return power to the grid for credit and receive a full loan under the so-called net

metering scheme for that.

- You can calculate the number of solar panels by creating a professional solar system or using the solar system size calculator.

Inverters are the devices for converting the solar panel Direct Current (DC) into the Alternating Current that homeowners require. There are three types of inverters:

- String: Least expensive, but inefficient, centralized inverters.
- Micro-inverter: Expensive and fixed on every solar panel, enabling even when some panels are shaded, to work smoothly.
- Power Optimizers: These are direct DCs for conversion to AC, mounted in each row. Less expensive, but slightly more than the string type as micro-inverters.

Panels are not necessarily connected to the wall. Panels are fixed to the roof and positioned to allow maximum sun visibility. They are installed in racks.

A solar monitoring system will show the house owner how much energy is produced per hour to check the performance of your photovoltaic system. The system can detect possible changes in performance.

Solar batteries can be installed for later or overnight storage of energy. Alternatively, net metering is available in some communities, which makes excess energy available for credit to the grid. Mainly, you will use the grid to store the surplus. It's like installing a solar battery at no cost.

How Much Is The Solar System Going To Cost?

The right system for any residential application can be evaluated, as stated in several variables. Understand, because the professional installation needs time, training, skills, and materials, the equipment cannot be the most costly component. Be mindful that while a standard refund can take between 6-7 years, certain businesses offer a 25-year guarantee. Consider the amount of service for at least 20 years and what you can save.

And keep in mind in computing that, while some states offer additional incentives, the IRS offers tax credit substantially for your panel system.

Who Should I Use For The Home Installation Of My Solar System?

Proper preparation and expenditure returns can be as important as the efficiency of the equipment for future results. Extensive training is essential, and every element needs to be fully understood. The best manufacturers have identified and worked with qualified installers in each geographic area. These companies offer training and inspections from professional on-site installers before certification.

What Is The Best Solar System For Families?

While many manufacturers believe they are the 'best,' one way to check for the best choice is to use comparisons from reputable sources. Two key elements to consider for long-term value are a) efficiency and b) warranty period.

Efficiency is a determining factor in how the sun's solar energy is converted into electricity per square foot. According to several solar websites, including SolarReviews, SunPower solar systems are leaders in the efficiency of solar panels. Still, their solar panels are also sold at a significant premium compared to other excellent solar panels that are almost as efficient.

The duration of the warranty depends on the manufacturer's trust in his products and the certainty that you are insured in case something happens. Many companies offer only a 5 or 10-year product warranty and a 25-year performance warranty, while SunPower offers 25-year industry performance, labor, and parts protection and guaranteed high power.

To find out if a solar panel system is a worthwhile and viable investment, use the solar calculator to gather the information needed to make a decision.

COMPONENTS FOR YOUR SOLAR PANEL SYSTEM (PHOTOVOLTAIC)

Solar power systems are a popular choice among renewable energy options due to the relatively low maintenance requirements and long-service lives of many system components. Since there are no moving parts, and therefore, there is little chance of mechanical failure, most solar power systems will produce electricity for 30 years or more!

Although some small solar power systems are relatively easy to install, many people hire installers. Regardless of whether you plan to install a booth system yourself or have a system installed by a contractor, you will benefit from understanding and proper maintenance of each component of your system.

PHOTOVOLTAIC MODULES AKA SOLAR FIELDS, SOLAR ELECTRIC PANELS OR PV MODULES

Photovoltaic modules are referred to as solar modules or solar collectors. We will use the terms in this section as synonyms, although the 'photovoltaic module' is the most technically correct terminology.

Solar panels supply electricity from sunlight. They generally comprise of silicon crystal wafers, which are referred to as cells, glass, polymer substrates, and aluminum frames. Solar panels can vary by type, size, shape, and color. In most cases, the 'size' of a photovoltaic module refers to the panel's nominal power or the potential for power generation. Solar panels also have voltage values. Those with 12 or 24 volts are generally preferred for off-grid systems with battery banks. Other solar modules have less common nominal voltages such as 18, 42, and even 60 volts. These modules are typically used in grid-connected applications to allow grid-connected inverters to function. The solar modules can be combined individually or in arrays by connecting them or inside to meet the required specifications. The price for most large PV modules for home or commercial use can range from $2.20 to $3.40 per nominal watt.

COMPENSATE THE SYSTEM (BOS)

In the terminology of photovoltaic systems, everything except the photovoltaic modules themselves is referred to as 'Balance of System' or BOS. We will analyze the following main components of the BOS one at a time in the direction of the flow of current through a typical system.

MOUNTING SYSTEMS FOR SOLAR FIELD OR PV MODULE

Solar panel mounting systems include hardware that allows the array to be permanently attached to a roof, pole, or floor. These systems are usually made of aluminum and selected according to the specific model, number of modules in the range, and desired physical configuration. Solar modules can operate at cooler temperatures, and properly mounted airflow can be refreshed around the panels. Wind load is a facility consideration for all roles, and the proper construction and change of concrete bases for each postal support is of considerable significance. These are an alternative for polar mounting to improve power output by moving the array to sunlight as the sun travels through the sky. A solar tracker array produces more energy than a fixed array. In water pumping applications, trackers are widely used. The cost of a tracker can be considerable and is recommended for mechanically inclined people due to the possibility of failure. The cost of a mounting system varies with the numbers of modules and the form of mounting. The average cost for a fixed array is $250-$1,000, and for a solar tracker is over $2,000. Another estimate of the costs for assembling the rack is between $ 0.50 and $ 1.00 for the array's nominal power.

COMBINER BOX

A combo box is an often overlooked and yet essential part of most solar power systems. The combo box is an electrical housing with which several solar modules can be combined in parallel. For example, if you want to connect two 12 volt panels for your 12-volt system, connect the output of each panel directly to the combo box's connections. From the combo box, it is possible to lay only a positive and a negative wire (in the corresponding line) to the next system component, the charge controller. The combo box also includes standard fuses or circuit breakers. These boxes are generally suitable for outdoor use and can be placed directly next to the array of solar modules. Combo boxes typically cost between $ 80 and $ 140.

SOLAR CHARGERS

Every solar system with batteries should have a solar charge controller. A charge controller regulates the amount of electricity that photovoltaic modules put into a bank of batteries. Its main function is to prevent overcharging of batteries. However, the charge regulators also prevent electricity from flowing back from the battery bank into the photovoltaic system at night or on cloudy days and from discharging the battery bank.

PWM (Modulated pulse width) and MPPT (Tracking) are the two major types. PWM is older and is most frequently used in smaller solar power systems. Choose your battery and solar array-identical PWM charging controller. The system must also handle the complete solar array current safely (in nominal amps). MPPT charging controllers can monitor the full solar panel PowerPoint and have 10-25% more power than the same pair of PWM controllers. These are achieved by converting surplus voltage into usable electricity. The ability to accept higher voltage from the solar panel for the output of a low voltage bank is another feature of MPPT charging controllers. Usually, charge controllers cost $50-$750 according to scale, form, and characteristics.

BATTERIES FOR SOLAR ELECTRICAL SYSTEMS

Batteries store electricity chemically in renewable energy systems. They come in different voltages, but the most common variants are 6 volts and 12 volts. The three types of batteries most commonly used in RE systems are:

- Flooded lead-acid batteries (FLA)
- Batteries sealed with absorbed glass mat (AGM)
- Sealed gel cell batteries

Flooded lead-acid batteries are the cheapest option. They require maintenance that monitors tension, adds water, and occasionally maintains. Besides, FLA batteries discharge hydrogen with a strong charge, so they must be stored in a ventilated compartment. Due to the maintenance issues of FLAs, some people prefer sealed, maintenance-free batteries. Since they are sealed, they don't need to be sprayed, nor do they normally emit gas. AGM batteries cost more and are more sensitive to overload than FLAs. Gel cell batteries are similar to general assemblies in that they are also sealed and require no maintenance, but are usually the most expensive of the three types. The life of all types of batteries is measured in units of time and is directly related to the number of possible recharge cycles; the more batteries are discharged with each use, the fewer recharge cycles are obtained from them. Sealed batteries usually don't last as long as flooded batteries. Well-maintained FLAs can last up to ten years, while sealed batteries have a lifespan of nearly five years. Other factors to consider are that some of these batteries weigh over 200 pounds and can cost between $20 and $1200 depending on their capacity. Therefore, consider energy conservation needs very carefully in light of maintenance problems, weight, and costs. Planning five days of battery storage for your system may not be the best option!

SOLAR INVERTER

An inverter takes the batteries (DC) and converts them to (AC), which is used to operate the most common electrical loads. There are two main classes of inverters or independent units that can be connected to the network.

Off-grid inverters require batteries for storage. In particular, the inverters connected to the grid do not use batteries, and the inverters compatible with the grid can operate with or without batteries, depending on the design of the system. There is a wide range of inverter functions available suitable for different system requirements and situations. Some inverters have built-in AC chargers so they can charge batteries from the grid in bright sunlight. Inverters with integrated AC chargers can also be used in connection with fossil fuel-based generators to charge batteries or operate very large loads. Off-grid inverters intended for home use must have suitable conduits and accessories that include all live cables. Normally, inverters for the whole family are designed for a continuous power of at least 2000 watts. Some devices (compressors or other inductive loads) and many sensitive electronic devices (chargers, computers, stereos, etc.) do not work properly with the modified sine wave power. Off-grid inverters can cost between $100 and $3,000, depending on the size and type.

A straight inverter is connected directly to the grid without using batteries. With these inverters, the photovoltaic system also fails in the event of a power failure, to protect the service lines from injury caused by unexpected 'live' lines in the event of faults. A grid-compatible inverter can be connected to the grid. The batteries can be used, which allows the possibility of an emergency power supply in the event of a fault. Typically, grid-connected inverters also produce 2,000 watts or more and cost from $2,000 to $4,000.

DC and AC Separations.

No code-compliant system can live without interruption! The DC and AC disconnect switches of a photovoltaic system are manual switches that can cut off power to and from the inverter. Some inverters have disconnectors with switches integrated into their structure. Other systems use an integrated power panel to support the inverters and associated disconnections in an organized arrangement. In other cases, the appropriate disconnector must be purchased separately to work with an inverter. The disconnectors are used by service personnel or by authorized persons (firefighters/police/electricians) to prevent electricity from a renewable energy system from reaching the inverter. (Don't forget that most inverters have capacitors that can hold a deadly charge up to several minutes after the input power is turned off. See the inverter manual for safe access times.) Disconnection prevents the generated electricity exceeding the current and disconnects the point from a faulty electrical network or a damaged component. Homeowners or authorized personnel can use an automatic switch to turn off a system for maintenance or assistance. The cost of disconnections can range from $100 to $300.

DIFFERENT COMPONENTS

Cables, plugs, lines, and brackets This category contains everything needed to connect all parts securely. As with most specialized technologies, many parts and tools are needed to properly install a safe and effective PV system. It is the installer's responsibility to understand exactly these and all rules and regulations for solar power systems (Section 690 of NEC is the key here). Acquiring the knowledge necessary to develop and install a safe and efficient system will ensure that your system meets your needs effectively. Still, it will also protect you and your home and promote the acceptance of renewable energy as a traditional energy source.

We are glad that you are managing a photovoltaic system. Regardless of whether you install the system yourself or hire a professional installer, you must adhere to the National Electric Code for security reasons. Our photovoltaic installation books and instructional videos also offer excellent references.

THE REALITY OF A SECURITY SYSTEM

We receive many questions about emergency power options in the home or office in the event of a power outage. Often these questions come from people in areas hit by hurricanes and where the power supply can be cut off for a long period. 'Backup power' can mean different things to different people. Therefore, we will try to explain some of the considerations in choosing the components for your system.

SOLAR ELECTRICAL SYSTEM LINKED BY UTILITY

If you plan to install a solar power system to compensate for electricity bills, use a system that turns electricity back on to the main network. If the power is interrupted, the system no longer feeds energy into the network and is no longer in the house. If you want to connect to the mains and continue to produce and use your power supply in the event of a fault, add at least some batteries and related components to the system. Below we help you understand the considerations for determining the appropriate battery backup amount for your system.

DRIVE A REFRIGERATOR - ONLY FOR A SHORT TIME!

If your electricity fails, you will want to make sure you spend hundreds of dollars on groceries in your refrigerator and consider a small all-in-one backup system. For about $300, a Xantrex PowerPack X1500 battery and a combination of inverters can power a medium-sized refrigerator for 2-4 hours before the battery needs to be charged. The AC adapter can be charged by connecting it to an electrical outlet. If prolonged outages occur, the power supply battery needs to be charged differently, e.g. from a solar panel. You can use a small 12-volt solar panel (between $30 and $350) to charge the battery in the power adapter. For example, a 60-watt solar panel would be exposed to full sunlight for about 3-5 days for a full charge.

ALL THE LOADS OF THE HOUSE

If you need to run multiple loads for several days while the power is turned off, you need to deal with a more expensive solution that requires more analysis. To size a large backup system, first test your execution, create a load list. Don't list everything in the house. just choose what you need to use during the break. This keeps costs low and can reduce system complexity. For example, you can install a system that powers your refrigerator, compact fluorescent lights, a fan, and a small TV.

You can use the debit list to determine the total amount needed on a typical day in the event of a power outage. Next, determine the number of days you think you don't have the sun to charge the batteries. This is also known as the days of autonomy. After a typical storm, the sun can emerge with full force. In other cases, it may not be possible to charge the batteries for a day or two. If you continue to consume energy from the batteries without recharging them, the charge will run out, and, worse still, the batteries may be damaged.

With the total watt-hour requirements and the desired number of days of autonomy, it is now possible to calculate both the size of the battery bank and the number of solar modules required for charging. You will find ideas for the components you need for the backup system in our off-grid living systems.

HOW TO SIZE A SOLAR SYSTEM: A STEP-BY-STEP WALKTHROUGH

After establishing that a grid-connected solar system is the best option for your home, we would like to help you properly size the system. In this section, you will learn how to size a solar system that covers your energy consumption patterns without oversizing your photovoltaic system.

As a systems designer, I follow a step-by-step process to size systems based on the dimensions that work with the constraints of my client's project.

The first step is to discover the main constraints for the project and use these constraints as a starting point for the design. We can approach the project from one of three angles:

- Budget constraints: Create a system within the set budget.
- Space restrictions: Create a system as space-saving as possible.
- Energy offset: Create a system that balances a certain percentage of your energy consumption.

I want to make sure that I provide a system that meets my client's specifications, but I also need to consider the size factors that may not be immediately apparent to them.

Some common obstacles that keep popping up:

- Local sunlight
- Array alignment (tilt angle)
- Plans for future expansions
- Evaluation of product efficiency
- Natural reduction in benefits during the warranty period

Taking into account the restrictions mentioned above, this section is intended to provide a detailed overview of the sizing process for grid-connected solar systems.

Sizing of photovoltaic systems connected to the grid: quick estimate,

First, determine the kilowatt-hour (kWh) consumption from the electricity bill. We would like to be available every 12 months to view the peaks and valleys used. Energy consumption increases in summer and winter when air conditioning and heating devices are used heavily. A whole year of data on energy consumption gives us the overview we need.

We would like to average the 12-month invoice data to determine the average monthly consumption of kWh. Networked systems tend to produce an excess in summer with maximum solar radiation.

If the service company offers a convenient guideline for net measurement, the system's energy can be registered with the service company as credit for later use. Not all utilities do this. Check with your local supplier.

Next, we would like to search for your sunshine hours per day using a sunshine hours chart or the PV watt calculator (I will come back to this in the next steps).

We can use this simple formula for a general estimate and then refine it as we move on to design:

(Annual kWh consumption ÷ 365 days ÷ average hours of sunshine) x efficiency 1.15 = size of the required solar DC system.

If the solar system cannot be exposed to the south at the preferred angle, it is necessary to adjust it by adding more solar energy.

Here is an example I live in New Mexico, where the photovoltaic watt calculator says I have an average of 6.10 hours of sunshine per day. It's a lot, I know, but that's why I live here. I use 1000 kWh per month or 12,000 kWh per year. According to the formula:

(12,000 kWh ÷ 365 days ÷ 6.1 hours of sunshine) x 1.15 = 6.2 kW DC solar system required

Project development of the estimated system

I search the address in Google Maps when I am ready to estimate the solar power system as precisely as possible. I'm checking if I have viable options for a south-facing roof bracket.

(Your solar power system should be facing the equator, so if you live in the southern hemisphere, look for options north instead.)

A roof bracket is a simplest and cheapest solution. It costs less than other racks. The roof pitch is often already designed for solar gain and keeps the solar modules close to the inverter and the service panel. This is excellent for efficiency and costs less in pipes and wires.

If a roof bracket is out of the question, I will consider the possibility of floor mounting or a solution mounted on a pole.

As soon as we know how much space we have for the solar panels and the angles and directions we will work, I take the photovoltaic watt calculator and follow these steps.

How to Use the PVWatts Calculator

- Enter the address and click the orange arrow on the right.
- If you are on the System Information page, enter the size of the CC system from the previous section.
- Select the standard module.
- For the array type, select 'fixed' for the roof brackets or 'open' for the floor brackets.
- Leave the system losses at around 15%.
- Enter the roof slope in degrees and the azimuth. Azimuth is the degree of north and south, where north is zero and south is 180. (Click here to find out how to optimize the angle and azimuth values.)

After entering all the information, click the arrow on the right to see how much electricity your system uses each month.

This is our step-by-step process to refine a system of precise dimensions. We provide this information because our audience is very do-it-yourself, and most people prefer to do their research at their own pace.

Once you are ready, we advise you to organize a free design consultation with us so that you can check the dimensions, find compatible products and make sure that the system works within your limits (budget, space, and energy offset). You can also call us at 1-800-472-1142 for immediate advice.

Selection of Solar Grid Systems

Once we know how big the solar system must be, we will compare it to the available space. If you are doing a floor mount, this is usually not a problem.

From my example above, I know I need a 6.2 kW DC system. I can multiply that number by 1,000 to confirm that I need 6,200 watts of solar panels.

My fastest resource is to go to our grid-connected solar packets and scroll down until I see something in this area. If the customer expresses the desire to purchase panels made in America or needs certain functions such as monitoring individual panels, I consider these options.

Here are some viable options that I would consider. Note that imported panels are cheaper, so you get around 10% more production at the same price.

Mesh binding systems with panels made in America:

- 6.2 kW system with Mission 310 W solar modules and SolarEdge inverters/optimizers
- 6.2 kW system with Mission 310 W solar modules and Enphase IQ7 + micro inverter
- 6.2 kW system with Mission 310 W solar modules and SMA central inverters

Grid systems with imported panels:

- 6.7 kW system with 335 W Astronergy solar modules and SolarEdge inverters/optimizers
- 6.7 kW system with 335 W Astronergy solar modules and Enphase IQ7 + micro inverter
- 6.7 kW system with 335 W Astronergy solar modules and SMA central inverters.

EIGHT TIPS FOR BUYING SOLAR MODULES FOR RESIDENTIAL BUILDINGS

In recent years, more and more companies have entered the solar module market.

These companies range from landscape architecture companies to television or electronics retailers who all have different levels of knowledge of the solar industry itself.

The purchase of solar panels is an important decision to invest that experts ought to trust. Not everyone creates solar panels the same. After several years of use, one panel can last for decades, and another can collapse. A series of factors, some of which cannot easily be seen by the finished panel itself, determine the length of installation of the solar panel.

Here are some of the many factors that can affect a panel machine quality:

- • Guarantees for renewable energy projected in the plan, with 100% solar generation for 25 years.
- • All the questions are frankly answered, regardless of whether you order from us.
- • No sellers are intrusive. We believe that it is our task to

educate you to manage the mess so that you can decide for yourself whether solar is right for you.

- • Solar panels, inverters, and shelves of high quality. We mount in our homes, just the same tools.
- • We will not propose until all systems are truly at home. All systems are customized. The internet tip doesn't threaten and doesn't alter.
- • Averages 6-8 weeks from signing to electrifying the solar system for quick installation.

You need to upgrade your solar panels every few years without the correct material combination for your device, unnecessary energy, money, and longer-lasting effort. Given the complexity of the solar purchasing experience, it is important to understand the topic and ask the right questions when making this major purchase.

These are eight tips you should consider when purchasing solar panels:

1. Keep a brand.

Today, there are many small manufacturing plants and cheap producers selling solar panels abroad. Even if something goes wrong, these smaller outfits cannot be avoided without a production error or bankruptcy. You have a system that has broken and no solution. It is better to stick to the listed firms or major US businesses for this reason. Herstellers can replace production problems or solve them and provide system guarantees.

2. This shows the PTC rating.

In the measurement of the solar panel system performance, two standard classifications must be taken into account: STC classification (standard test conditions) and PTC classification (PVUSA test conditions). The STC testing is carried out in the plant, and the PTC testing is a PV power measurement in the 'real world' under testing conditions. The larger the ratings, the more effective the panel becomes. To ensure maximum system performance, you should aim for a PTC/STC ratio of at least 90%.

3. Verify the negative rating of tolerance.

A negative index of resistance is a measuring device that does not adhere to manufacturing standards for the solar panel. The assurance deal may be impacted. Unless the client has a high level of adverse resistance in his deal, he may not be able to do what he considers to be a flawed device. The manufacturer would consider this 'normal,' for example, if it had only been obtained from the Solar Modul for 180 watts using a 200watt solar module with a 10% tolerance negative. Try shooting a panel system with a 3% or less negative response.

4. Get a guarantee.

Make sure you receive a warranty on your system for at least 25 years to ensure your solar module's life expectancy. Demand the written guarantee of a manufacturer rather than a dealer warranty.

5. Look for panels with high inverter efficiency.

Inverter performance is the system's ability to produce energy that impacts your product portfolio directly. With a system with low inverter efficiency, energy loss over the system, life will cost thousands of dollars. Ensure you are looking for a solar module with an inverter efficiency of at least 95.5% to maximize your investment and ask for CEC weighted efficiency information rather than a full image of the top inverter efficiency.

6. Evaluate the design and quality of the mounting system.

Your panel mounting system needs heavy aluminum rails that can stand up to extreme conditions, regardless of the climatic environment. Small or inexpensive mounting systems will easily be damaged and take time or money unnecessarily.

Make sure the production system uses aluminum extrusion at least 6105-T5 and provides a 10-year warranty.

7. Calculate the price per watt.

Calculating costs per watt can be a reasonable solution to ensure that you get the best value for your money with two similar devices. If two systems contain the same solar modules, mounting systems, and inverters, calculating the price per watt can be a good deciding factor. You shouldn't pay more than $3.94 / watt (equipment only) or $5.50 / watt (installed) for a standard 4kW string inverter system.

8. Module efficiency.

This criterion must only be taken into consideration in situations where there is little space for solar modules. The higher the efficiency of the module, the smaller the panel. While 'higher module efficiency' sounds great, the fact that a smaller but more efficient solar panel may not provide enough energy to power your home compared to a larger module. Besides, models with higher efficiency tend to have higher costs for the same performance. If you have enough space for the solar panel system, why should you pay thousands more to save a few inches?

FACTORS TO CONSIDER WHEN BUYING SOLAR MODULES

If there is one completely free thing, it is sunlight. It doesn't matter who you are, what you do ... the sun will shine equally for all of us. But the best thing is how we can use it as energy. The sun shines in the sky and wants to be used forever. It is time for you to take advantage of it. Because infinite free energy is a dream come true. You only need solar panels for this. Since there are many options, you need to be careful when choosing.

Here Are Five Factors To Consider When Choosing Solar Modules.

1. The purpose of the purchase

The functionality of each solar module differs from the other. This simply means that not all solar panels are created equal and that not everything fits everything. It is necessary to ensure that the selected panels meet the specific requirements. For example, solar panels that work for a restaurant may not work in a typical home. Likewise, matching and buying by purpose are crucial.

2. The size of the place where it is to be installed

Sometimes the type of position or surface on which the panels are to be installed is neither uniform nor typical. In such cases, shapes and sizes need to be adjusted as needed. If such a situation has occurred, contact your service provider.

3. The expected services

The amount of energy consumed by different locations in the same type of premises is different. For example, the energy required for a one-story house is different from the energy required for a three-story house. The scale is important. As long as you have selected an expert dealer, they will be the best to advise you.

4. The warranty period

You shouldn't buy any type of solar panel that isn't covered under warranty, no matter how it seemed. The reason the warranty period of solar modules is quite long is that they should not break easily. Therefore, it is better to invest in a quality product.

5. The reputation of the dealer

Some corrupt dealers intentionally provided you with defective items and blame you. For this reason, you need to do your research, read customer testimonials, and make a selection. Because the quality of the products depends on it.

SOLAR PANEL INSTALLATION MANUAL - STEP BY STEP

Tips for installing the solar panel

The questions to be asked before mounting a solar panel are:

Where can I buy solar panels?

Solar panels can be purchased from various solar companies and even from online stores.

Where are the solar panels installed?

Solar panels are generally installed on separate roofs, buildings, or structures. It is important to install the solar panel in a place most directly exposed to the sun.

Solar panels work with optimal performance in direct sunlight. When installing the solar power system, try to place the photovoltaic modules directly under the midday sun to achieve maximum efficiency of the photovoltaic system.

Beware of obstacles to sunlight before installation. Remove any unnecessary obstacles and objects such as branches that could block the sunlight of the solar unit. You should also follow the path of the sun in the sky and make sure that no objects cast a shadow on your solar photovoltaic modules. The operational efficiency of your solar system suffers from this shadow.

Types of supports for solar panels

Photovoltaic modules are installed with brackets for solar panels. These brackets are available in 3 main types:

1. Tree supports
2. Brackets for attic
3. Rinse the brackets

With the help of these brackets, you can install your solar panel on a camper, on the roof or the side of a pole on the roof. You can even install it as a standalone unit.

Things to do before installing the solar panel

Cost calculation

The first step is to calculate the cost of setting the type and size of the system. Keep in mind that the government of various countries around the world offers subsidies to promote renewable energy installation through solar panels. This subsidy is different in different countries. For example, the subsidies offered by the United States differ from those in India or China.

Equipment needed

The second step is to create a checklist of devices needed for a solar system: solar modules, charge controllers, inverters, and batteries.

System size

The next step is to determine the size of the required solar system. You need to add the power of all the electrical devices you want to use. Calculate how many hours per day devices are used.

If you follow the previous steps, you can determine the power requirements, the required solar battery size, and the cable size. Remember that the correct cable sizes prevent overheating of the cables and ensure maximum energy transfer to the batteries.

Solar panel installation manual - step by step

Solar panels can be used to generate electricity for both commercial and private use. In both cases, the photovoltaic modules are installed on the roof to receive the maximum possible sunlight and generate the maximum electricity from the system.

The steps of the installation process are as follows:

Step 1: Install the assembly.

The first step is to repair the brackets that support the solar modules. Depending on the needs, these can be attic brackets or flush brackets. This basic structure offers support and robustness. Attention is paid to the direction in which the photovoltaic modules (monocrystalline or polycrystalline) are installed. The best direction for solar panels is the south for the countries of the Northern Hemisphere as they receive the highest sunlight. The east and west directions are also sufficient. For the countries of the southern hemisphere, the north is the best direction.

Here too, the assembly structure must be slightly inclined. The angle of inclination can be between 18 and 36 degrees. To maximize conversion performance, many businesses use solar trackers.

Step 2: Install the solar panels.

The next step is the connection of the solar modules to the assembly structure. This is done by tightening screws and nuts. It takes care to adequately protect the entire structure so that it is robust and durable.

Step 3: Make the electrical connections.

Electric wiring is the next step. Wiring, common connectors like MC4 are used since they can be connected to all sorts of solar modules. The following sequence can be attached electrically to these panels:

1. Sequence connection: In this case, one PV module 's positive (+) wire is attached to the other module 's negative (-) cable. This type of wiring increases the voltage regulation on the battery bank.

2. Parallel connection: In this case, a positive (+) to positive (+) and negative (-) to negative (-) connection is established. This type of wiring voltage for each panel remains the same.

Step 4: Connect the system to the solar inverter.

The next step is to connect the system to a solar inverter. The positive cable from the solar panel is connected to the positive pole of the inverter and the negative cable to the inverter's negative pole.

The solar inverter is then connected to the solar battery and the power supply to the grid.

Step-5: Connect solar inverter and solar batteries.

In the next step, the solar inverter and the solar panel are connected. The positive pole of the battery is connected to the positive pole of the inverter and from minus to minus. A battery in the off-grid solar system is required to conserve emergency power.

Step 6: Connect the solar inverter to the electricity grid.

The next step is to connect the inverter to the grid. To make this connection, a normal socket is used to connect the main control panel. An output cable is connected to an electrical circuit that supplies electricity to the house.

Step: 7: Start solar inverter

When all the electrical connections have been made, it is time to start the inverter switch on the main switch of the house. Most solar inverters have digital displays that show statistics on the generation and use of solar systems.

5 WAYS TO AVOID SHORT CIRCUITS

Short circuits are a serious type of electrical accident that can seriously damage the electrical system. They occur when a low resistance path unsuitable for the transportation of electricity receives high electrical current. In simple terms, short circuits occur when the hot wire touches a conductive object; it shouldn't.

The result of a short circuit can be damage to equipment, electric shock, or even a fire. And if you don't take preventive measures against short circuits, you only increase the risk of these situations occurring. Roman Electric recommends that all homeowners in Milwaukee practice short circuit prevention practice.

The following are 5 of these steps.

1. Check the outlets before use.

There is a box with wires connected behind each socket. Some of the main causes of a short circuit are defective cables, loose box connections, and an obsolete electrical outlet. While diagnosing these problems can be difficult because they are hidden behind the walls, it is still possible to avoid short circuits by examining the electrical outlets before each use. There are some indications of the risk of a short circuit in the socket:

- The outlet has burn marks or a burnt smell.
- Hum or crackle from the socket.
- Sparks are emerging from the outlet.
- The outlet is over 15-25 years old.

If any of these guidelines apply, do not use the electrical outlet and contact Roman Electric immediately.

2. Check the devices before use.

As with electrical outlets, devices must also be checked before connecting them. Short circuits can also be caused by incorrect cables or circuits from the device itself. Check the device for the following signs before each use:

- Damaged cables, housings, or wires.
- Several cracks in the device.

The device has exposed circuits.

If you apply one of these signs, we recommend that you dispose of the device or have it repaired by a specialist.

3. Reduce electricity consumption during thunderstorms.

One of the most dangerous ways a short circuit can occur is lightning, as the overwhelming amount of electricity can cause serious damage. It is recommended to reduce energy consumption only during a thunderstorm. This not only prevents short circuits during a storm but also reduces damage when a surge occurs.

4. Perform basic circuit breaker maintenance.

Your electrical system is protected against short circuits. These are its automatic switches. These components in the control panel are turned off when electrical currents are classified as unstable, each of which is connected to a different circuit. We recommend that you perform some basic maintenance on the circuit breaker to make sure they work.

Here are some tips:

- Check each circuit breaker for damage, cracks, or loose parts.
- Know which circuit controls each switch. It is recommended to use an automatic switch detector. Further information can be found here.
- Clean dirt or stains on the switch and panel (use only a dry cloth).

If you wish to improve the maintenance of circuit breakers with professional services, contact Roman Electric to obtain our circuit breaker services.

5. Plan an electrical inspection at least once a year.

Similar to a doctor's appointment, electrical inspections should be performed at least once a year. This way, an electrician like Roman Electric can fully examine your electrical system. From there, we can locate and prevent short circuits from occurring and offer convenient solutions to solve other problems. Electrical inspections help maintain wiring, electrical outlets, and any other part of the electrical system. Plan one today!

FIVE WAYS TO SAVE SOLAR ENERGY

With the ample ability to reduce your monthly bills, switching to solar is an excellent way to reduce the burden on your bank account.

No wonder the acceptance of solar modules in New Zealand is growing surprisingly rapidly. The total energy generated by solar panels in New Zealand in 2016 was estimated at 51.7 GWh, an increase of 52% over the previous 12 months! *

In this section, we examine five different ways you can make and save money instantly and in the long run by installing solar panels in your home.

1. The most obvious advantage of switching to solar is that it will lower your electricity bills every month. These savings save installation costs from the first day of installation. While the exact savings from one family to another seem slightly different, we have customers who regularly save 50% or more of their electricity bills every month. Some even went as far as saving up to 90% in the summer, making their home as energy efficient as possible with a little help from our secret tips and tricks.

2. Most New Zealand energy companies offer a repurchase price for families who use solar panels to generate excess energy. These

energy retailers offer the following (as of December 2016):

- Contact: 8 cents per unit

- Mercury: 8 cents per unit or 12 cents for newly supplied solar energy customers who conclude a 3-year contract

- Trust: 7 cents per unit

- Meridian: 8 cents per unit

- Genesis: 8 cents per unit

- Ecotricity: 7-8 cents per unit

- HP power: 16 cents per unit for the first 50 kWh, which are exported every two weeks, then 8 cents

3. Solar panels very often add value to your home. As an example of an overseas (but very relevant) example, a study conducted in the United States from 2003 to 2010 in San Diego and Sacramento found that solar panels add an average $20,000 to the value of the real estate.

4. Heating costs can be reduced in winter by connecting solar panels to the heating system. With the heat contributing significantly to electricity bills in most New Zealand households, this is a great way to cut down on these expensive winter costs.

5. At World Solar, we offer battery storage systems. This stores

excess electricity which can be collected and used in low light conditions. This means that the energy you consume will never be wasted.

Solar energy offers several important financial benefits that are beneficial to both your bank and the environment.

The five benefits described above quickly compensate for installation costs from the first day of installation and offer clean and environmentally friendly renewable energy that can save you high costs in both the short and long term.

ENERGY-SAVING TIPS THAT WILL ALSO SAVE YOU MONEY

Saving on electricity bills is a commonly overlooked way to cut down on expenses. Reducing only 20% of monthly energy bills can make a big difference in the budget, especially when considering the average monthly electricity bill in the United States at almost $120.

Fortunately, there are countless ways to reduce energy consumption. Read on for our best tips, and you will find something that works for you.

Replace your bulbs with energy-saving options.

Incandescent lamps waste a lot of electricity. It is difficult to justify them if you can choose between LED, CFL, or halogen lamps in a range of colors, temperatures, lumens, and prices suited to your needs. Although they are more expensive than ordinary incandescent lamps, they pay many times more while saving on electricity costs.

Use a surge protector to not drain energy from your device's chargers.

When the phone charger is connected, it consumes electricity, even if the phone does not charge. To avoid wasting energy, connect your phone, laptop, and other chargers to a surge protector that can be turned off when not in use. When it no longer charges, you can disconnect all devices from the power supply with the touch of a button. And remember; these energy-absorbing devices also weigh on your bank account!

Turn on the grill.

Would you rather grill your summer evenings outdoors or sweat on a hot stove? It seems to be an easy answer. If you grill outdoors, you can enjoy the fresh air and not take the heat indoors. Even better, you save a lot of money on cooling costs.

Use your ceiling fan.

With a ceiling fan, a room can feel a few degrees colder, especially when used in conjunction with the air conditioner. Remember ceiling fans, nice people, not rooms. Therefore, make sure fans are switched off as you leave the rooms.

Turn the switch.

A large number of roof fans can send air down (to refresh) or up (to heat). If you want to feel warmer, make sure that the ceiling fan blades revolve in the opposite direction, as it will contribute to the cooling air. Once done, you can turn down the thermostat, open the windows, and enjoy the fresh air. Of course, you can also enjoy energy saving!

Dust it all.

When we say all the dust, we mean it. These include ventilation slots, light bulbs, ceiling fan blades, baseboards, electronics, and office equipment. Because? They all consume more energy when covered in dust. Tip: Open the doors and windows while dusting to make sure they are gone forever (or at least a little longer).

Vacuum the vents of the dryer and the cooling coils.

Like the rest of the electronics, the dryer and refrigerator work more efficiently when they are free of dust. For this purpose, the dust may not be sufficient to completely clean the air vents of the dryer and the cooling coils. Therefore, use a suction hose fitting to thoroughly clean these devices. This will allow the refrigerator and dryer to run smoothly.

Reduce the temperature settings on the water heater.

Did you know that you could waste energy because the water heater thermostat is set too high? That's right: the higher the temperature, the more energy the water requires to boil. To save energy, set the thermostat of your water heater to a temperature below 140 ° C (higher and mix only cold water).

Automate your thermostat.

Regulate the temperature at night and when no one is at home a few degrees lower. While you can do it without a programmable thermostat, let's face it; it's hard to remember doing it constantly every day. By switching to a 'set and forget' model, you easily save on energy costs.

Move lamps, televisions, and other potential heat sources away from the thermostat.

While rearranging all the living room furniture, consider moving heat sources such as lamps and TVs away from the thermostat. Because? If they are too close together, this can cause the thermostat to believe that it is a few degrees warmer than it is. And this keeps your air conditioning system longer - and wastes electricity and money. If you have a company, turn down the thermostat before your guests arrive. Collective body heat also increases room temperature.

Become solar.

If you are not solar powered yet, you should consider this. Switching to solar energy can save you between $7,000 and $30,000 in 20 years, depending on where you live. Not only do they guarantee low electricity tariffs for the life of solar modules, but they also support a local and sustainable form of clean energy.

To save money and the environment follow these tips.

As you can see, you can save a lot on energy charges to make your home more energy efficient. The best part? Many of these energy-saving tips don't cost you a cent. It is a win-win situation: you save money and at the same time, protect the environment.

CONCLUSION

If you are one of those people who are concerned about the environment, you know that other forms of electricity cause a lot of pollution.

While others may think that pollution from the use of electricity is inevitable, this concept has long been disproved by the invention of solar panels and solar cells for power generation. Solar energy for families not only protects our environment but also saves long-term money with solar energy.

Before continuing to discuss solar energy and everything related to it. First, let's discuss what solar energy is for the home and how it works.

Household solar energy is a power system in which people can use electricity for their homes using solar energy. This means that people can illuminate their homes, cook food, heat water and space, watch TV, and use equipment and devices only with the sun's energy.

How does it work? Solar energy for domestic use is used by using passive heat, which is responsible for heating our homes and water, or by using photovoltaic cells, which are responsible for the supply of light and the functionality of our devices.

While solar energy at home has many benefits for humanity and the environment, the production cost of these modules and solar cells is quite high due to the type of material used. Therefore, this becomes a problem for people who want to use solar energy for their homes because not all people can afford the necessary amount.

Because of this need, it is a positive step forward that there are solar plans for private homes on the market. With solar home plans, people can create their solar panels without having to pay so much. However, some beginners want to know what to look for in a solar home plan, so they don't waste their money.

Here are some tips you can use:

A solar home plan must include a list of the necessary materials and the tools needed to build your home solar energy. It should include a list of stores or suppliers that indicate where to buy the necessary materials and order them based on the suppliers that offer the best deal.

The plan must also include detailed instructions for carrying out the activities. If a video is included, it is all the better because it systematically guides you.

Plans should also include a follow-up that can be used to determine how much energy is produced and how much energy is consumed. It should also contain storage information to conserve the excess energy you need on rainy or cloudy days.

With this information, you can now easily check the right solar plan for your home. You will be surprised that the whole process is fairly simple and fun.

www.ingramcontent.com/pod-product-compliance
Lightning Source LLC
Chambersburg PA
CBHW070233220526
45465CB00004B/1417